LARGE-SCALE INCIDENT MANAGEMENT

A Small Town Plan
for a Big City Problem

Mark Haraway

DELMAR
CENGAGE Learning™

Australia • Brazil • Japan • Korea • Mexico • Singapore • Spain • United Kingdom • United States

DELMAR
CENGAGE Learning™

Large-Scale Incident Management
Mark Haraway

Vice President, Career and
Professional Editorial:
Dave Garza

Director of Learning Solutions:
Sandy Clark

Acquisitions Editor:
Janet Maker

Managing Editor:
Larry Main

Senior Product Manager:
Michelle R. Cannistraci

Editorial Assistant:
Maria Conto

Vice President, Career and
Professional Marketing:
Jennifer McAvey

Marketing Director:
Deborah S. Yarnell

Marketing Manager:
Erin Coffin

Marketing Coordinator:
Shanna Gibbs

Production Director:
Wendy Troeger

Production Manager:
Mark Bernard

Senior Content Project Manager:
Jennifer Hanley

Art Director:
B. Casey

Technology Project Manager:
Joe Pliss

Library of Congress Control Number: 2008930890

ISBN-13: 978-142835993-2

ISBN-10: 1-42835993-1

Delmar
5 Maxwell Drive
Clifton Park, NY 12065-2919
USA

Cengage Learning products are represented in Canada by Nelson Education, Ltd.

For your lifelong learning solutions, visit **delmar.cengage.com**

Visit our corporate website at **cengage.com.**

Notice to the Reader

Printed in Canada
1 2 3 4 5 XX 10 09 08

∼ Dedication ∼

I would like to dedicate this book to my wife, Sheila, who has served as my typist, editor, and overall supporter throughout the writing of this book. But more than this she has been with me for the last 21 years through every fire, emergency, and other departmental event. She has always supported my career and kept the household running while I was serving my community. She is what made this book possible; she kept me going and made sure that I stuck to the project.

I also want to dedicate this book to my children, Joshua and Jacob, whose patience and perseverance has remained a constant force in my life. Josh, thank you for your volunteer fire service on the most challenging fire of my career. It is an honor that you have chosen the fire service as your career. And thank you, Jacob, for never questioning why I had to be gone, only asking when was I coming home.

CONTENTS

PREFACE

On August 23, 2005, a storm system named Katrina formed off the coast of the Bahamas. This system was headed for the Gulf Coast of the United States. On August 29, 2005, Katrina struck the Gulf Coast with intense ferocity. The wind and rain that both preceded and accompanied this storm as it made landfall presented a larger, more catastrophic incident than could have ever been imagined. I am sure we can all still remember watching as floodwaters swallowed up New Orleans. As the citizens of New Orleans pleaded for assistance, and while the neighboring state of Mississippi was all but leveled along its coastal plain, the federal government's response was less than desirable. This is evident from President George W. Bush, who stated during a news conference on September 15, 2005, that "four years after the frightening experience of September 11, Americans have every right to expect a more effective response in a time of emergency. When the federal government fails to meet such an obligation, I, as President, am responsible for the problem, and the solution." President Bush went on to order a comprehensive review of the federal response so that we as a nation could make the necessary changes to be "better prepared for any challenge of nature or act of evil men that could threaten our people."

I use this reference to get you to think: If we as a country can identify the need to review and enhance our plan to provide for a stronger response to any event whether natural or man-made, then it stands to reason that we as local first responders may need to do likewise for the protection of the citizens we serve. As you read forward, you will learn of another event that occurred October 5, 2006, that was not a product of Mother Nature or a terrorist attack. The event was man-made—caused by the same chemicals that are used to make many of the products that we use in our daily lives. It threw a small community into the forefront of disaster. But because of preparation, planning, and training, this town's responders, supported by numerous state and federal agencies, were able to meet the challenges of this disaster head-on and successfully mitigate the event.

This book will describe the event in detail so you can understand how to perform a local hazard assessment for your community, develop a response plan, and structure the management of a large-scale incident. I leave you with the following quote from Max Mayfield, Director of the National Hurricane Center: "Preparation through education is less costly than learning through tragedy."

How to Use This Book

This book was designed to be used both as an educational text and a resource manual for first responders. It is different from other texts written on the topic of incident command and incident management because it is based on an actual event that occurred in Apex, North Carolina, which it references throughout the chapters so that the reader can actually see how the Incident Command System and the concept of incident management can be used successfully. To use the book effectively, start at Chapter 1, which gives the background of the Apex event, and then go forward to develop an understanding of large-scale incident management.

The text will review the Incident Command System. After this review, you will read about developing a hazard assessment for your community. You will learn how to take the information from this assessment and apply it to the development of an Emergency Operations Plan, (there is even a sample plan in the NIMS Resource CD). Then, the book takes you through the planning process, and you learn how to plan for your event.

Next, the book takes you to the event: What do you do now? Don't worry—that is covered too. You will learn how to manage resources, implement the planning process into your event's organizational structure, and develop your management plan.

Supplemental CD

The book comes to you with a comprehensive supplemental CD that includes numerous NIMS documents, lists Web-based resources, and offers the reader two PowerPoint presentations. One presentation is designed as a 12-hour educational program that takes students through the concepts of large-scale incident management and then places the student in a simulation that requires the use of his or her management skills to run an emergency event. There is even a student handout provided detailing the events of the Apex incident.

The CD provided with this book includes the following:

- A 45-minute PowerPoint presentation of the Environmental Quality Chemical Fire and the emergency response with a section on lessons learned.
- A 12-hour instructional PowerPoint that follows the book and is designed to be used for teaching a class on large-scale incident management.

- A sample Hurricane Plan that can be used as a guide to developing or enhancing your own Emergency Operations Plan. This plan includes a section on hazardous materials, search and rescue, and post-event recovery. (This is included in the book as Appendix A.)
- A sample Family Action Plan designed to be used by on-scene personnel before the event. The Family Action Plan identifies what the first responder should do personally prior to an event and also how to ensure that his or her family is ready. This is important since the first responder is generally activated to respond before or during an event, therefore, leaving family members to fend for themselves. This plan also includes a contact sheet that lists emergency numbers for the family and where they will be. This allows the event General Staff to verify wellness of responders' families so that the responders can carry out their assigned tasks without worry. (This is included in the book as Appendix C.)
- A sample Emergency Pet Plan for use in handling pet and animal needs during an emergency. (This is included in the book as Appendix D.)
- Summations of Presidential Directives HSPD 5 and 8 for reference.
- An instructional spreadsheet that identifies training requirements as set forth by NIMS.
- Actual Apex Fire Department timeline of the event as it unfolds.
- Blank Incident Management Forms in both PDF and Word formats for emergency responders to use for practice or on the job.

ACKNOWLEDGMENTS

I would like to thank the following agencies and individuals:

Town of Apex Fire Department

Career and volunteer firefighters and administrative staff whom without your support and dedication daily to this fire department this book and my job would be impossible.

Town of Apex Government Staff

Mayor Keith Weatherly, Manager Bruce Radford, Police Chief Jack Lewis, EMS Chief Nicky Winstead, and all town employees who pull together every day to get a job done in the highest standards possible.

Janet Maker and Michelle R. Cannistraci at Delmar, Cengage Learning who gave me this wonderful opportunity.

Robin Reed and the editorial staff at S4Carlisle whose patience and guidance helped bring this project together successfully.

Fire Chief Lee Barbee of the Clayton Fire Department and other reviewers whose feedback was invaluable to this project.

Reviewers:

Lee Barbee
Fire Chief, Town of Clayton
Clayton, NC 27528

Andy Byrnes
Program Developer
Emergency Services Dept.
UVSC
Provo, UT

Rick Haase
Conoco Phillips Wood Refinary
Roxana, IL

Alan Joos

Stephan S. Malley
Weatherford College
Weatherford, TX

Mark Haraway is the Fire Chief of Apex Fire Department in Apex, North Carolina. He is a third-generation public safety responder with 29 years of dedicated service to fire, rescue, and EMS, serving the last 6 years in the Town of Apex Fire Department.

Chief Haraway graduated summa cum laude with a bachelor of science degree in fire safety management. He holds a diploma from the University of North Carolina's Institute of Government in Public Administration and is a graduate of the Executive Development Program for North Carolina Association of Fire Chiefs. He holds many teaching certifications and has developed his own programs such as "Large Scale Incident Management" and "Protecting Your Own." He sat on the development board for the NC State Fire Marshal's Office for Unified Command and Control and the first Confined Space Rescue Technician Program.

Chief Haraway is currently a certified Fire Instructor for seven North Carolina community colleges as well as a Certified Rescue Instructor with specialties in trench rescue, confined space rescue and hazardous materials. He serves as a Representative of North Carolina Urban Search and Rescue Technical Advisory Board, a Representative for Wake County Fire Service Level Committee, past Member of the NC Hazardous Materials Advisory Board, past Treasurer of the NC Fallen Fire Fighters Foundation. Chief Haraway is an Adjunct Instructor for NASA Ames Research Center in California as a Structural Collapse Specialist. He served as a delegate for the National Council on Readiness and Preparedness (NCORP) as well as a certified CHS III with the American Board of Certification in Homeland Security. Through a cooperative partnership, he developed a Technical Rescue Team to become a part of North Carolina Task Force (NCTF) 4, serving as a Team Leader for NCEM deployment to Western North Carolina with NCTF 4 in response to post-hurricane flooding, and was one of the recipients of the Langley Higgins Award for Swift Water Rescue in 2005.

INTRODUCTION

The story you are about to read in Chapter 1 is true. It is a story of a small town thrown into a potentially disastrous situation and how, through planning and training, emergency responders were able to respond and successfully handle it so that there were no environmental repercussions and no fatalities. For centuries, everything from war time battles to world's fairs to celebrity gatherings have been planned out to the most finite detail. It is time that we put the same care and attention into planning to respond to man-made and natural events that affect our communities.

The most effective way to enforce this is through the use of the National Incident Management System (NIMS) and the Incident Command System as referenced in NIMS. All local and state government agencies as well as tribal governments have been directed to implement training and procedures in line with NIMS. However, many are hesitant to believe in its effectiveness. Many look back at events such as Katrina and are concerned that the system will not work as it was designed. Well, the story you are about to read will reveal how the system should and can work when used properly.

The event is the Environmental Quality Chemical Fire of 2006 in Apex, North Carolina. This event alone set new benchmarks as to how agencies should respond to large-scale events that affect entire communities and locales. Sandia Labs, the United States Fire Administration, and the United States Chemical Safety Board have studied the event.

As you read the story, put yourself in the different roles as an emergency responder. What would you do? Would you know what to do? Does your department or community have a plan to respond to an event such as this, a weather event, or any type of large event that may tax your department, government, and community resources? Understand that it can happen anywhere, and no one is immune. That is why after reading Chapter 1, you must read on through the rest of the book to capture the behind-the-scenes story—how Apex had a plan, trained its personnel to utilize that plan, and executed that plan to manage and respond to this event. This book looks at how to develop your city or town's Emergency Plan, how to perform a pre-incident survey, what to do once the event occurs, and how to complete an Incident Action Plan.

The other thing that you will begin to understand as you read on is the responsibility placed on the Incident Commander (IC), Command, and

General Staff. Do not forget that the IC is responsible for everything not delegated. But in the large-scale event, all eyes are on the IC; personnel, government leaders, civilians, department heads, and event state and federal agency representatives look to the IC for answers. You, as the IC, regardless of your occupation will be expected to make the decisions and direct the operations necessary to mitigate the situation. Your actions will have to be well orchestrated, and operations will have to run like well-oiled machines. Your actions will be scrutinized, and you will need to be prepared to stand behind and, at many times, defend your decisions. It is important that you remember this and use the material in this book to help you develop a response plan that will protect your community.

OVERVIEW OF THE INCIDENT COMMAND SYSTEM

> *"Self confidence is the first requisite to great undertakings."*
> *Samuel Johnson*

Learning Objectives

After reading this chapter you should be able to

- Understand the Incident Command System
- Define responsibilities of Incident Command Staff
- Define responsibilities of General Staff
- Define chain of command, unity of command, and unified command

This introductory section is provided as a refresher on the Incident Command System for those individuals who may not, as part of their normally assigned duties, be responsible for managing an emergency situation. In this section, we will review key components of the Incident Command System from its origin to how the system is used today as a standard for managing incidents and events on a daily basis.

Origin of Incident Command

The Incident Command System (ICS) originated in California in the 1970s. The California Department of Forestry was instrumental in the development of the system for use as a management tool for the many large wildland fires it had to deal with. What California found out was that the Incident Command System gave it the management tools it needed to control the large number of resources that it was deploying to combat its fires. As the system developed, it became obvious to many leaders within the structural firefighting arena that this same system could be used for the management of structural firefighting events and could be expanded further as an all hazard management tool used to effectively run any type of event whether natural or man-made. Phoenix Fire Department's retired Fire Chief Alan

Brunacini identified this and expounded upon the idea with the development of his book *Fire Command*. *Fire Command* introduced the concept of incident command to the fire service to be used as a daily management tool applicable to any size of incident.

Today, government agencies are required to use the Incident Command System on all events. This is based on the requirements as specified in Presidential Directive 5, which prescribes training requirements under the Incident Command System for federal, state, local, and tribal government groups and their personnel.

Attributes of ICS

From this point forward, the Incident Command System will be referred to as ICS. Let's first look at ICS as an incident management system. In doing so, we need to identify the attributes of the system. ICS at a routine event is usually simple, and the trained responder can use it with relatively few problems. One reason for this is that on the routine event, there is a limited number of resources, the Incident Commander can oversee all necessary management functions, and the event is confined to a single jurisdiction or responding authority. However, for a complex event requiring a larger management organization, the ICS must be able to expand and grow to meet the needs of the event. Some cues to this need for expansion are: problems rapidly multiplying, the need for an increased number of resources to handle the event, and the need for an expanded staff to address specific areas such as safety or media relations.

The ICS is designed to expand in a modular fashion and thus be organizationally structured to address specific incident needs. You should also remember that span of control is a key component of how this expansion will unfold. Span of control is generally defined as the number of individuals or resources that a supervisor can effectively oversee. Typically, span of control is three to seven, with its optimal value being five. This number may represent individuals or single resources. In the next section, we will break down this modular expansion, identifying functional positions and their responsibilities. Then we will define the Command and General Staff positions.

Role of Incident Commander

The Incident Commander (IC) is the individual who is responsible for management of the overall event. The IC may be the first arriving law enforcement officer or the company officer on the first arriving fire apparatus. That is the first basic principle of the system that needs to be understood. ICS is not based on departmental rank and file; it is designed so that first arriving units can initiate the system and start the incident management process. The role of the IC can change as the event evolves and other ranking personnel arrive. However, remember that command is only transferred after a face-to-face meeting has been held to discuss the status of the event. The incoming IC will need to review what has transpired and what needs to occur next. The following are the IC's responsibilities:

- Determine strategy to resolve the event.
- Select tactics to be employed for incident stabilization.
- Develop an Incident Action Plan (IAP); the IAP will be discussed in detail later in this book.
- Create an effective ICS organization for effective scene control and incident management.
- Manage incoming resources.
- Coordinate and identify resource activities.
- Provide for scene and personnel safety.
- Identify a methodology for the media release of information about the event.
- Coordinate relations with all outside agencies responding to the event.
- Coordinate release and demobilization of incident resources.

As you look at this list, you may say that this is too much for one person to do and still be responsible for the situation at hand. Well, in some situations you may be right. But think back to the last routine event you responded to. It is likely that you completed all these tasks and did not even realize it. How? During routine responses such as a "room and contents fire" or "automobile accident with minor injuries," first responders automatically implement basic incident management skills. However, on large-scale incidents requiring a multi-agency response or events that affect multiple jurisdictions your ICS structure will need to expand to address the many responsibilities that Command will be facing. This is where you expand and implement Command Staff positions.

Command Staff Positions

By assigning Command Staff positions, the IC will be able to maintain a manageable span of control. The Command Staff positions assigned under the IC will take care of incident safety concerns; media and community information issues, which include press releases related to the event; and serve as a liaison to other agencies who are on-scene and have information relevant to the situation. Next are the staff positions that may be assigned under the IC and a description of their responsibilities.

Safety Officer

The Safety Officer is the individual responsible for on-scene safety and the safe operations of responding personnel. The IC is responsible for the duties of this position until it is staffed. This position is normally staffed at a point that the IC realizes he or she cannot oversee the safety of responders, perform on-scene safety analysis, or if the event involves hazardous materials (OSHA 1910.120 requires a safety officer on all events involving hazardous materials). The Safety Officer is responsible for overall scene safety. Therefore, this individual will have to have some prerequisite knowledge of the job. As an example, you would not place a rookie in the role of Safety Officer nor would you want a firefighter overseeing the safety of a law enforcement event. Remember, the safety officer has the authority to take immediate action if imminent danger is identified or personnel are operating in an unsafe manner. If this happens, Command must be notified.

Liaison Officer

The Liaison Officer is responsible for serving as an interface with other responding agencies that are not involved directly with operations, but may offer supporting information. Examples would be the American Red Cross or public utilities. This position would be staffed as the situation unfolds, requiring supporting agencies to respond and coordinate with the IC. If the IC cannot effectively work directly with these agencies, then the position of Liaison Officer should be filled. If this position is staffed, then a Liaison Area should be established adjacent to the Command Post for receiving assisting individuals. By having a separate area, the IC will be able to focus on the identified strategies of the event.

Information Officer

The position of Information Officer should be staffed once the media arrives on the scene and the need for event information is established. The individual serving in this role will need to be well versed on public etiquette and how to communicate effectively with the public. The Incident

Command Staff should provide frequent updates to the Information Officer and approve all informational releases. You will find that if you establish an area for the media and other agencies requiring information, the release of information will be more manageable. Understand that this position can make or break your event. The public must be kept informed. If you provide the necessary information to keep the media informed, your event will be perceived as under control. The public will feel they know what is going on in their community. Media resources can also be of assistance to you, especially in the event of an evacuation or any other need that has to be relayed to the community in response to your event. This is why you should build a relationship with them before the event.

Creating the General Staff

As an event expands, your incident management system will expand not only in the Command Staff positions, but also in the operational arena under General Staff positions. The General Staff positions are staffed as the IC identifies the need to implement tactical actions to stabilize the event. The General Staff positions are Operations, Planning, Logistics, and Finance. Next, we will identify what each position is responsible for. Individuals serving in General Staff roles will be identified as Section Chiefs.

Operations Section Chief

The Operations Section Chief is the individual responsible for managing operational resources assigned to the event, determining and directing tactical operations, allocating and assigning resources to bring stabilization to the incident, as well as participating in the development of the Incident Action Plan. This position should be staffed as the event expands or requires more resources, and Command finds it necessary to concentrate on strategy and not be involved in tactical operations. Once this position is staffed, Command focuses on the "big picture," defines the necessary strategies required to bring the incident under control, and works to develop necessary components of the Incident Action Plan.

Planning Section Chief

The Planning Section Chief is the individual responsible for collecting and evaluating information that will be used to develop the Incident Action Plan. This individual will record resource status, document the incident, and aid in the development of the Incident Action Plan.

Logistics Section Chief

The Logistics Section Chief is the individual responsible for providing facilities, services, and materials to support the incident. This position will identify and provide the necessary resources as they relate to service needs and support functions. The Logistics Section Chief is responsible for providing resources that support operational personnel working at the incident such as food, shelter, medical assistance, and hygiene requirements.

Finance/Administration Section Chief

The Finance/Administration Section Chief is the individual responsible for all financial requirements and cost documentation of the event. This position should always be staffed when the event expands to a point where unusual costs are incurred or there is a financial responsibility to the responding jurisdiction that must be addressed and documented accordingly. The individual assigned to this position will have to be given the authority to make any necessary decisions related to the financial aspects of the incident.

Unit

This is an organizational element having functional responsibilities for specific incident activities in Planning, Logistics, and Finance.

All of the aforementioned General Staff report directly to the IC. If the positions are not filled, then the IC will be responsible for the identified duties of that position.

Expanding the General Staff

ICS allows for the General Staff positions to expand for larger incidents. Under this section, we will look at the use of Branch Managers, groups and divisions, crews, task forces, and strike teams.

Before continue, I must define what a *single resource* is. A single resource is the term used to reference an individual company of personnel with an apparatus or an established team of individuals with an identified work supervisor. The following is a list that identifies specific ways that single resources can be used to better organize an expanding workforce under the Operations Section.

1. **Crew**—This is a specific number of personnel without an apparatus. A Crew is assembled for an assignment and will have a designated leader.

Generally, the number of personnel will be based on the normal span of control, which is five to one. The Crew will be identified based on the assigned task, such as "Ventilation Crew" or "Attack Crew" and such. A Crew would be used just as any other "single resource."

2. **Task Force**—A Task Force is any combination of single resources that are put together to carry out an assignment. This Task Force assignment will be one that is temporary such as exposure protection or something that once it is completed the Task Force will return to staging for another assignment. A Task Force will be supervised by the Task Force Leader. Once again, span of control should be used to determine the number of resources assigned to a Task Force. An example of a Task Force would be a team made up of two engine companies, two ladder companies, and a rescue truck.

3. **Strike Team**—Specific combinations of the same kind and type of resources, with common communications and a designated leader.

4. **Division**—A Division is used when the incident is divided geographically. If the Division is assigned to handle an outside ground task, their identification will be alphabetical. However, if the Division is operating in a structure or building, their reference is based on designation related to the floor of the assignment.

5. **Groups**—These are teams of personnel assigned by their functional tasks. A Group is assigned based on the needs of the incident. Groups and Divisions are equal in authority; however, a Group is not confined to a single geographic area. It is the Group's responsibility to carry out its assigned functional tasks, wherever that may be within the area of the event. Division and Group Supervisors must work together and communicate when working in close proximity to one another, but neither holds authority over one another.

6. **Branch**—The Branch is an organizational level having functional, geographical and/or jurisdictional responsibilities for incident operations. This level generally falls between the Section Chief and Divisions or Groups if assigned to the Operations Section, and between Units and Sections in Logistics. Branches can be identified in many ways to include their function, jurisdictional name, or with the use of Roman numerals.

Refer to Figure I-1 for an example of an ICS organization from the NFA ICS 300 student manual.

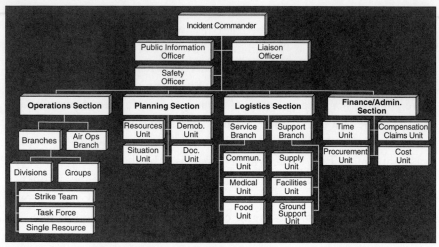

Figure I-1 *ICS organization. Courtesy of the USFA.*

Previously, we reviewed the basics of organizational structure as it relates to the ICS. Remember, the ICS is used on all incidents. It is designed to expand and contract with the incident.

Communications and the Chain of Command

One of the key things to ensure effective use of the ICS is communications. The first thing to learn in good communications is the "chain of command." The **chain of command** is defined as an orderly line of authority and reporting between the different levels of the organizational structure with a smooth flow from lower levels to higher levels of the organization (see Figure I-2). The chain of command aids in the communication of direction to subordinates and maintains control by management. However, the chain of command should not interfere with the sharing of information. Orders always follow the chain of command, but the exchange of information may be done on an individual basis when requested. The chain of command also does not always follow the organizational rank and file of day-to-day administrative roles. For example, just because a Chief Officer of the department shows up on the scene, they do not have to take the role of IC. In fact, he or she may be assigned a subordinate duty based on the needs of the incident.

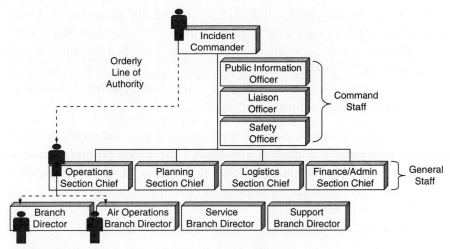

Figure I-2 *Chain of command. Courtesy of the USFA.*

Unity of Command

Unity of command is also an important term to remember. This is not to be confused with the term unified command. Unity of command is the concept that everyone reports to one supervisor, and workers on the scene receive assignments from only their supervisors. Unity of command ensures that assigned personnel report only to their assigned supervisor and receive direction from that same supervisor. Unified command is a team effort that allows all agencies with jurisdictional responsibility for the incident to manage the incident by establishing a common set of incident objectives and strategies.

Common Terminology

It is very important to use common terminology and clear text for radio communications. When outside resources arrive on large-scale events, they will need to know where to report and receive assignments. Without common terminology, no two agencies will be able to work jointly together without confusion. The same stands true for clear text radio communications.

Ten codes were a good concept; however, they do just what they were designed to do when used on a large-scale event where multiple agencies are on-scene, insures secrecy through radio security. There are so many variations to the 10-code list that seldom are ever the same. Due to these variations, agencies have a difficult time determining which code is correct. With clear text, everyone understands that a fire is a fire and a bomb is a bomb—no confusion.

In the communications arena, you will have two types of communications with personnel, defined as follows:

- Formal—communications with personnel to give and receive assignments, request additional resources, and make progress reports. Formal communications require that orders, directives, resource requests, and status changes must follow the hierarchy of command unless otherwise directed.
- Informal—used to exchange information about an incident or event. This is important because critical information must flow freely between personnel.

Good communication is imperative to the management of your event or incident. Command and General Staff must have an orderly and effective transfer of information related to the event. Whether written or verbal, effective communication will be paramount in determining how your incident will end.

Modular Organization

Remember that the ICS organization is designed to be modular, expanding as necessary for the type, size, scope, and complexity of the incident. Your ICS organization should build from the top down. As the need is identified, you can add sections that in turn will have subordinate units as required. Your organizational expansion should match the identified functions and/or tasks that need to be performed. You should add staff only as needed to accomplish specific tasks or complete functional elements of your organization. Under this same concept, if you do not assign the tasks to an organizational element, they will be completed by the next highest level of responsibility in the organization. Once tasks/functions are completed, those organizational elements can be deactivated. For the most part, there are no set rules for the expansion of your ICS organization. You should only fill necessary positions; for every position that is implemented, someone must be in charge to oversee that function. This modular expansion concept is what makes the ICS so applicable to the emergency services as well as other organizations responsible for managing task-based projects. From the smallest event to the largest incident, you can apply the management practices of ICS to effectively coordinate your activities.

Chapter Summary

In this Introduction, you have read about the basics of the Incident Command System. From its introduction in the 1970s to today's requirements for it to be used as a daily management tool for all activities by federal, state, and local government agencies, the ICS system has proven itself as a formidable management tool. This Introduction gave the reader an opportunity to review the basic ICS structure as well as how the system expands and contracts with the size of your situation. Good communication is key to ensuring that tasks are completed and assigned personnel understand their assigned roles. As you continue to read the following chapters, the ICS will be discussed in even greater detail, presenting the many designated roles and responsibilities that can be assigned.

Review Questions

1. Who is responsible for determining the strategy necessary to resolve the event?
2. Which OSHA standard requires that a Safety Officer be assigned to any event involving hazardous materials?
3. Who is responsible for the management of all operational resources?
4. True or False? The Planning Section Chief is responsible for collecting and evaluating information to be used in the development of the incident IAP.
5. A concept used in incident command that states that everyone reports to only one supervisor is called _____.
   A. unified command
   B. unity of command
   C. supervisory control

Chapter 1

Strategic Planning for the Large Incident

> *"Good tactics can save even the worst strategy. Bad tactics will destroy even the best strategy."*
>
> General George S. Patton Jr.

Learning Objectives

After reading this chapter, you should be able to

- Identify the benefits of having an All Hazard Emergency Operations Plan
- Define National Incident Management System and Incident Command System
- Identify how NIMS and ICS were used to bring the Apex fire to a successful end

Responding to the Large Incident

The problem of how to handle and respond to the "large incident" has always existed. Any time a severe or catastrophic emergency situation occurs, all eyes turn to the first responder to develop a solution and implement that solution. In the end, the expectation is that everything will be the same as it was before the event. Unfortunately, many times that is not the case. There are many reasons for this less than desired outcome. Some will ask if the event was just too enormous for first responders to handle effectively. Others will ask if there was a plan in place; and question whether the individuals attempting to respond to the event had the proper training. It is my summation and observation that many times the correct answer would be "all of the above." This is not to be taken as a critical statement toward any previous response, nor do I mean that previous events with less than desirable outcomes could have been prevented. However, with proper training, the use of the

Incident Command System (ICS) and **National Incident Management System (NIMS),** and when the implementation of a plan that is practiced, catastrophic events can be managed in a very functional manner—thus leading to a more successful event.

I want to take you back to an event that set new benchmarks in the areas of evacuation, management, and response. The incident was the Environmental Quality Chemical Fire in Apex, North Carolina, that occurred on October 5, 2006. This event tested the skills of first responders and city officials alike. However, to fully understand the concept of how the proper use of **Incident Management** made this event end successfully, one must look at what was done prior to the event. This is because a true Incident Management Plan must be developed, written, practiced, and reviewed regularly to be functional. This same plan should be written to NIMS compliance and meet the definition of an **All Hazard Plan.**

Pre-Event Disaster Potential

The Town of Apex is a suburb of Raleigh, North Carolina, in Wake County. Apex is a rapidly growing town that has seen its population grow from 6,000 to more than 32,000 in the past seven years. A fire department working out of three stations with 27 personnel supported by 15 volunteers, a police department of 48 officers, and an EMS service of three ambulances provide emergency services for Apex. In Apex, due to limited staff, the fire chief acts as the chief fire official and the emergency management coordinator.

The potential for disaster in Apex had been identified much earlier than October 2006. In 2002, Apex suffered a severe blow from a winter storm that brought an inch of ice to the town, causing a major substation fire that left citizens powerless for nine hours. During the winter storm event, it became evident that the existing Town of Apex Emergency Plan was inadequate and outdated as a tool to assist responders in major events. Most of the emergency contact numbers were incorrect or no longer in use. There was no sheltering plan or identified sheltering location for the general population. The plan was not well organized and did not identify any type of command structure. This event led to the development of an **Emergency Operations Plan** and a **Sheltering Plan** to be used in response to major man-made or natural events that affected Apex. In May 2006, the plan was rewritten to meet current NIMS guidelines. Once this was complete, all departments were trained in accordance to the plan as well as to current Incident

Command practices as outlined in NIMS IS-100, *Introduction to Incident Management,* and IS-700, *Introduction to National Incident Management System.* The training was designed to be compatible with Wake County and other neighboring municipal agencies. The Emergency Operations Plan was written to parallel the existing Wake County Plan.

Disaster Strikes

With an understanding of the pre-event actions that had taken place prior to October 5, 2006, let's look at what occurred that evening at 9:38 p.m. Firefighters were summoned to investigate a chlorine odor in the area of Investment Boulevard and Schiefflin Road in Apex. This area is one of the original industrial corridors of the town and has several occupancies that store hazardous materials. One of those facilities is the Environmental Quality Company, which handles and stores hazardous waste. As first units arrived, we were met with a vapor cloud 75 feet tall extending throughout the industrial corridor and well into a neighboring residential area.

Immediately, firefighters made an attempt to identify the source of the vapor and requested assistance from numerous agencies. I assumed the role of **Incident Commander** and activated the countywide **Reverse 911 system,** which sends emergency messages out to identified areas in the event of a disaster. This event would soon fall well within that definition. As reconnaissance teams continued to assess the situation, Engine 1 identified the source as the warehouse at the Environmental Quality Company.

Before firefighters could advise **Command** of the source of the vapor release, the entire facility burst into flames, immediately followed by violent explosions sending toxic clouds of smoke and fire hundreds of feet into the night sky. See the blast over the horizon in Figure 1-1. Due to the hazard and contamination of the air surrounding the scene, only one picture of the building burning exists (see Figure 1-2). See Figures 1-3 and 1-4 for the aftermath of the explosion.

As units arrived to assist, everyone was instructed to relocate to a safer position, and the evacuation area was immediately extended to include both residential and business areas for a 22-block radius around the site. This would soon be extended to a mile in all directions based on current wind conditions. From the time of the first alarm at 9:38 p.m., the Command Post was relocated four times in 30 minutes, eventually ending up at a location one-quarter mile from the fire scene, where an initial stand would be made. Figure 1-5 shows roadblocks that were set up to contain the area.

Figure 1-1 *Firefighters watch the horizon as numerous fireballs can be seen from a quarter of a mile away. The sound from the explosions resembled that of artillery fire and would shake the ground with each blast. Courtesy of Lee Wilson, photographer.*

Figure 1-2 *This picture identifies the collapsing remains of the main warehouse, which is being devastated by fire and multiple explosions during the Environmental Quality fire. Courtesy of the* News and Observer.

Figure 1-3 *This photo identifies only a small portion of the chemical containers housed at the Environmental Quality warehouse. The barrels identify what first responders faced— burned out chemical containers with no labels and pressurized containers, such as the expanded barrel above, that had not yet exploded. Courtesy of the Apex Fire Department.*

Figure 1-4 *An aerial view of a portion of the Environmental Quality facility shows a collapsed warehouse and fires still burning 20 hours after the initial response. Courtesy of the Apex Fire Department.*

Figure 1-5   *Roadblocks were set up to contain the area. Courtesy of Mike Legeros, photographer.*

Managing the Evacuation

Town law enforcement officials were assisting with the evacuation, and local media was being used to notify citizens of the hazardous situation that was unfolding. A media site was identified adjacent to the Command Post and shelters were opened for fleeing citizens. Elected officials were briefed on the situation and what the initial plan was. I explained that due to the many unknown chemicals housed at the Environmental Quality Company, the firefighters would not be able to make any type of fire attack until morning. In addition, due to current weather conditions and projected wind patterns, City Hall, the first shelter site, fire stations 1 and 3, and the 911 center would need to be shut down. The **Exclusionary Zone** would be extended to the full one-mile radius to ensure citizen safety in all affected areas. Two more shelters would be opened, and the Red Cross would be contacted to assist with caring for sheltered victims.

At this point it became necessary to identify the required command and general staff positions to better manage the many resources that would be needed to mitigate this event. Dispatchers were moved to a mobile communications unit from the fire department to maintain emergency communications, and all phone exchanges were routed to the neighboring Wake-Raleigh 911 center. Law enforcement officers and medical personnel were dispatched to each shelter to assist any evacuees as needed and help maintain order. Media briefings were made every hour with emergency messages scrolling all local channels with information on **Sheltering in Place** and available shelters for evacuees. Pets were addressed at each shelter by the utilization of a **Pet Plan** included in the Apex Emergency Operations Plan. The Town of Apex Emergency Operations Plan has an annex that addresses how to shelter and deal with animals during an evacuation. As part of this annex, appropriate housing capabilities and safety concerns are addressed to assist shelter management teams with receiving family pets and ensuring their needs are met during an emergency event.

The Exclusionary Zone Expands

As the fire and vapor cloud grew, weather conditions began to change, forcing an expansion of the Exclusionary Zone, which encompassed a local nursing facility. Command requested that the **Medical Branch** develop an evacuation plan for this facility to include all patients and employees (see Figure 1-6). The plan called for 16 ambulances, 2 buses, and 3 engine companies to converge for an emergency movement of patients and staff. As this was being done, Command was working with **Logistics** to request more resources to include a second Hazardous Materials unit, North Carolina Air Quality, EPA, Department of Transportation officials, and the school board.

The town **Emergency Operations Center (EOC)** had been relocated with the loss of fire station 3, and Command advised the town manager to close all local government offices for October 6. Logistics notified public works to send heavy equipment and sand in preparation for containment operations. As air quality technicians arrived, they were assigned to work under the **Hazardous Materials Branch** to establish perimeter monitors that would allow Command to validate perimeter borders and any contaminants being released from the fire. As Command continued to monitor the situation, the decision was made to close the air space for a five-mile radius over the city as well as stopping all rail traffic along the CSX railroad line. This included rerouting the Amtrak service, which, in turn, required bussing passengers to the next closest stop.

Figure 1-6   *During the Environmental Quality Chemical fire, first responders were not only faced with community evacuations but also the evacuation of a 100-bed nursing facility. Courtesy of the Apex Fire Department.*

Weather a Strong Factor

Weather continued to affect the situation. At 3:30 a.m., meteorologists monitoring conditions around Apex contacted Command to advise that a low pressure system had developed west of town that would change the wind direction 180 degrees by 5:30 a.m. and bring as much as 2.5 inches of rain to the area. Command called for a staff meeting to develop a fifth relocation plan of all on-scene resources. The new site would be 2.2 miles from the fire. As Command was being relocated, first responders, many of which were police officers, started showing signs of chemical exposure. The signs included tearing eyes, running nose, reddening and rash developing on the skin, nausea, and vomiting.

The Hazardous Materials Branch Director was told to work closely with EMS to establish emergency decontamination and transport of all exposed responders. A total of 12 police officers and 1 firefighter were transported to local medical facilities for treatment. Command briefed the manager and elected officials of the responders' condition. With the relocation drawing near and the situation not getting any better, Command met with staff to discuss the **Incident Action Plan (IAP)** for the next 12 hours. The discussion centered on school closures, school bus routing, extended sheltering plans, continued public information, and maintenance of the perimeter. All local hospitals were updated on the situation, and mutual aid units were dispatched to set up decontamination sites at all receiving hospitals in the event of any citizens self-deploying to hospitals for care. At 5:30 a.m., the alternate EOC and Command Post were relocated to the new Command Post, which was 2.2 miles from the incident, and a new media location was established to continue to provide information to the public. Joint information lines were established with the Wake County EOC and Wake County Public Health.

At 10:00 a.m. on October 6, initial reconnaissance was performed to identify the extent of damage and the existing fire conditions. Due to the size of the response to this event and the distance from the site, an **Incident Complex** was created, which allowed for a second **Operations Section** to be established for the purpose of allowing the hazardous materials unit to move into the exclusionary zone, creating a forward work area and decontamination site. This was set up at a local high school only a mile from the fire. As this process was taking place, Command was meeting with air quality officials to review perimeter monitoring results and any contaminants that had been identified. Command also met with Environmental Quality officials and arranged for an industrial firefighting team to be brought in to assist with the firefighting effort. In Figure 1-7, the firefighting team enters the fire scene to determine the size of the remaining fires.

The Fire Is Contained

After 12 hours, firefighters contained and extinguished the fires by using firefighting foam agents. The Hazardous Materials Operations Section had used the heavy equipment and sand provided by public works and created an earthen berm around the site perimeter prior to the firefighting effort, thus reducing runoff. With the fire extinguished, Command met with on-scene toxicologists from the North Carolina State Air Quality Office, Environmental Protection Agency, and the Centers for Disease Control and Prevention and determined that the outer exclusionary zone was safe for reentry. Command met with the Law Enforcement Branch Director and after developing a phased reentry plan, transferred Command to the police chief to oversee the reentry of citizens.

The Command structure remained a **Unified Command** with Law Enforcement serving as the lead agency during reentry. The fire chief, still

Figure 1-7   *Firefighting team entering the facility to evaluate the remaining fire situation. Courtesy of the Apex Fire Department.*

a member of the Unified Command, continued to work with the Hazardous Materials Section to monitor conditions and, with the event stabilized, assigned the **Planning Section Chief** to work with Logistics to develop a Demobilization Plan.

The Outcome

After 44 hours, Fire Command was terminated at 5:00 p.m. on October 7, 2006. Law enforcement units remained on the scene for almost six weeks post-event providing security at the site during cleanup efforts. After environmental engineers and toxicologists determined that the affected area was safe for reentry, all displaced citizens were allowed to reenter the exclusionary zone, and the town was repopulated. All affected businesses and public buildings were reopened, and life in Apex started to return to normal.

The results of this event are as follows:

- 17,000 citizens evacuated in three hours with no fatalities
- 100 patients moved from a nursing facility with no injuries
- 3 shelters established with a Pet Plan in place

- 9 schools closed
- Continuity of local government shut down, but all services remained intact
- Loss of two fire stations, police station, EMS facility, and 911 center, yet not a single call missed
- 35 victims seen at local hospitals to include 13 first responders with no overnight stay or extended health effects
- 252,000 air samples and perimeter samples taken during event
- To date no identified contamination to soil, water, or air

How did Apex accomplish this? Was it sheer luck? Did a good outcome just happen? Or was the outcome of this event brought on by the implementation of an effective plan and training compliant with the NIMS? As you continue through this book, you will develop an understanding of the use of the ICS and how NIMS can be used to manage any type of situation, from a small event faced daily in emergency services to a major event such as the one in Apex.

You will learn how effective planning in your town or city is the first step. We will discuss pre-incident planning and why it is the first step to developing an Emergency Operations Plan. To get started, you will need to survey your response area, identifying the high hazard areas such as railroads, chemical facilities, large industrial complexes, and high traffic routes. You may be in a rural area; if this is the case, don't feel that you are without risk. In rural areas, concerns can arise from applications such as spraying operations and silos. And we should all remember that Mother Nature does not discriminate—a fire or natural weather event can occur anywhere and bring with it catastrophic damage.

Looking Forward

Once you complete your planning survey, you will be introduced to the many types of Incident Management as outlined under NIMS through the use of the ICS. We will look at command and general staff positions and how this modular management system is capable of expanding with the incident. You may already be using ICS within your department. If you are, then you are ahead of the game. However, most emergency services organizations only utilize ICS as a single resource tool.

The following chapters will expound ICS and define Multi-agency Command, Unified Command, Incident Complex, and Area Command.

We will define what an Emergency Operations Center is and how it is utilized within your management structure. We will also discuss how to use your planning survey to develop an All Hazard Emergency Operations Plan, including all responsible agencies, identifying their roles in specific types of events. Within each segment of the plan, you will need to use materials from the NIMS/ICS chapters to develop an organization chart for each agency. Each agency will need to identify its responsibilities during events and create an objective-based chart to outline what role it will play during the event. By doing this you will develop a plan that any member of the agency can implement. And since it is written as an All Hazard Plan, it should address any potential situations that you have identified during your survey. The question then arises: What happens if the plan does not address a specific situation? Well, that is the best thing about having a plan. There is no way you can address every possible thing that may happen. What you will learn in Chapters 3 and 4 is how to identify each department's roles and responsibilities. The plan may not have a specific section, such as "How to Respond to a Plane Crash," but you have identified who is responsible for certain applications within your response area. You have included an organizational chart for each responding agency in your town or city, and you have an objective-based outline of what each department or agency will do to include what resources they can provide. With this information, you can apply the plan to whatever occurs and use it to bring your event to successful fruition.

The Big Picture

This book will walk you through how to survey your area, create a plan, train to use the plan, and effectively implement the plan when needed. The benefits of the ICS management concept, as outlined in NIMS, is that it does not have to only be used for large-scale emergencies; it can be used to manage everything from the local Christmas parade to a downtown festival. Using your plan for these events will give agencies an opportunity to prepare for the big event. That is what happened in the Apex event. Apex initially rewrote its plan after an ice storm in 2002, and then in 2006 added the necessary information to make it NIMS compliant. Apex then trained all employees on the effective use of NIMS and the ICS management concept. During a 54-hour window in time from October 5 to October 7, 2006, the hard work paid off, and what could have been a catastrophic event ended in success.

Chapter Summary

This chapter offered key terms and definitions for NIMS and the ICS as well as the use of an All Hazard Plan and how planning, training, and implementation were instrumental in a potentially catastrophic event. This chapter summarized that event and reviewed some details as to how strategic planning and training were used.

Key Terms

Incident Command System
 (ICS)
National Incident Management
 System (NIMS)
Incident Management
All Hazard Plan
Emergency Operations Plan
Sheltering Plan
Incident Commander
Reverse 911 System
Command
Exclusionary Zone

Sheltering in Place
Pet Plan
Medical Branch
Logistics
Emergency Operations Center
 (EOC)
Hazardous Materials Branch
Incident Action Plan
Incident Complex
Operations Section
Unified Command
Planning Section Chief

Review Questions

1. What are three key steps that will aid in the management of an emergency event?
2. What management system has been in place and used by emergency responders since the early 1970s?
3. What event caused Apex to rethink their Emergency Operations Plan?
4. Due to the magnitude of the Apex fire and the need for multiple agencies to take certain responsible roles, _____ Command was used to create a management structure that would place all responsible parties in the Command Post.
5. Define Pre-Incident Planning.

Chapter 2

Expanding on the Incident Command System

Learning Objectives

After reading this chapter, you should be able to

- Define Major/Complex Incidents
- Define Incidents of National Significance
- Classify Major incidents by type
- Define Span of Control
- Identify General Staff and Support positions
- Understand how to expand the basic principals of the ICS

Major/Complex Incidents

Major incidents equate to only 10 percent of the total incidents that occur each year. This may be one reason why many local jurisdictions do not focus on them and are therefore sometimes ill-prepared to respond to them effectively and efficiently. What is a **Major/Complex Incident?** It is an event that has one or more of the following criteria:

- Involves more than one agency and/or political jurisdiction.
- Involves complex management and communications issues.
- Requires experienced, highly qualified supervisory personnel.
- Requires numerous tactical and support resources.
- Involves multiple victims with injuries, fatalities, and/or illnesses.

Major/Complex Incidents also generally include one or more of the following characteristics:

- Widespread environmental or property damage.
- Psychological threat and trauma to the victims and the affected community.

- Lasts many days, thus resulting in the need for multiple operational periods.
- Are very costly with regard to control and mitigation.
- Require extensive recovery efforts.
- Draw national media coverage.
- Defined as an Incident of National Significance.
- Requires the management of donations and activities of nongovernmental organizations.

An **Incident of National Significance** is an incident so overwhelming for the resources of state and local authorities, that they have to request federal assistance to help with control and mitigation efforts. Examples of such incidents are as follows:

- Major disasters or emergencies as defined under the Stafford Act.
- Catastrophic incidents whether natural or man-made, including terrorism, that result in extraordinary levels of mass casualty, damage or disruption of services affecting the population, infrastructure, environment, economy, national morale and/or government functions.
- Events involving more than one federal department or agency during the control efforts.
- Events in which a federal agency, in response, has requested assistance from the Secretary of Homeland Security.
- Events where the Secretary of Homeland Security has been directed to assume responsibility for the management of the event by the President.

The aforementioned information is based on definitions as listed in the National Response Plan and as set forth in Presidential Directive 5.

Type Classification of Events: Defined by the United States Fire Administration (USFA)

Incident Typing is another way to classify major or other types of events. The concept of incident typing was developed as part of the NIMS. Incident typing is based on numerous criteria that includes the number of resources necessary to respond effectively to an event and the number of operational periods. The typing complexity ranges from Type 1 to Type 5, with the Type 1 incident being the most complex and resource intensive. Generally, Type 1, 2, and 3 incidents are considered to be "major or complex." However, you must take into consideration that a Type 3 incident in one jurisdiction may be considered routine in another simply due to available resources. Next are the defining characteristics of each incident type.

Type 5 Incident

- The incident can be handled with one or two single resources with up to six personnel.
- Other than the Incident Commander, Command and General Staff positions are not activated.
- There is no written Incident Action Plan required for this event.
- The incident is usually contained within an hour or two after resources arrive on the scene.
- Examples would include a vehicle fire, an injured person, or a police traffic stop.

Type 4 Incident

- Under this incident type, Command and General Staff functions are activated only if needed.
- Several resources are required to mitigate the incident, including a Task Force or Strike Team.
- The incident is typically contained within one operational period in the control phase, usually within a few hours after resources arrive on scene.
- The Agency Administrator may have briefings to ensure that the complexity analysis and delegation of authority is updated. This refers to the departmental manager or local government official being briefed to ensure that the nature of the event is properly typed and that the appropriate authority has been given to the Incident Commander for handling the situation.
- No written Incident Action Plan is required, but a documented operational briefing will be completed for all incoming resources.
- Examples may include a major structure fire, a multi-vehicle crash with multiple patients, an armed robbery, or a small hazmat spill.

Type 3 Incident

When capabilities exceed initial response capabilities, appropriate ICS positions should be added to match the complexity of the incident.

- Some or all of the Command and General Staff positions may be activated, as well as Division/Group Supervisor and/or Unit Leader positions.
- When managing the Type 3 incident, a Type 3 Incident Management Team or Incident Command Organization will manage the initial incident and be responsible for all response actions using a large number of resources. This same command organization will be responsible for extended response actions until the event is under control. Management under a Type 3 Incident Management Team will continue until the event transitions to a Type 1 or 2 Incident Management Team.
- The incident typically extends into multiple operational periods.
- A written Incident Action Plan is typically required for each operational period.
- Examples would include a tornado touchdown, earthquake, flood, or multi-day hostage/standoff situation.

Type 2 Incident

When the incident extends beyond the capabilities for local control, and the incident is expected to go into multiple operational periods, it becomes a Type 2 incident. It may require the response of out-of-area resources, including regional and/or national resources, to effectively manage the operations, command, and general staffing.

- Most or all of the Command and General Staff positions are filled.
- A written Incident Action Plan is required for each operational period.
- Many of the functional units are needed and staffed.
- Operations personnel normally do not exceed 200 per operational period, and total incident personnel do not exceed 500 (this is simply a guideline).
- The Agency Administrator is responsible for the incident complexity analysis, Agency Administrator briefings, and the written delegation of authority.
- Typically involves incidents of regional significance.
- Examples would include large wildfires affecting multiple jurisdictions, large moving weather events, or large hazardous materials events such as an oil spill.

Type 1 Incident

- This type of incident is the most complex, requiring national resources to safely and effectively manage and operate.

- All Command and General Staff positions are activated.
- Operations personnel often exceed 500 per operational period and total personnel will usually exceed 1,000.
- Branches may need to be established to aid in the management of General Staff Span of Control. Remember that under the concept of basic Incident Command, effective span of control is three to seven subordinates. To maintain this, the Incident Commander may deem it necessary to expand the organization by using Branches, Divisions, or Groups.
- The Agency Administrator will have briefings and ensure that the complexity analysis and delegation of authority are updated.
- Use of resource advisors at the incident base is recommended.
- There is a high impact on the local jurisdiction, requiring additional staff for office administrative and support functions.
- Typically involve incidents of national significance.
- Examples would include events such as the World Trade Center attacks or the attack on the Pentagon.

The previous information on incident typing comes from the USFA, which you can find on the included NIMS resource CD.

The Apex event expanded rapidly to a Type 3 incident and at the height of the response would have been classified as a Type 2 incident.

How Complex Incidents Arise

Generally, events are classified as complex in one of two ways. The first is when smaller incidents, such as a fire or hazardous materials spill, become major events as a result of wind or surface conditions or as a result of response delays, poor initial management, or lack of resource support. The second way occurs when the event starts out as a major event. Earthquakes, hurricanes, floods, aviation crashes, tanker spills, major hazardous materials releases, simultaneous civil disorders, or terrorism can all produce major/complex incident management situations. No geographic location is free from the potential for a major event that may require the use of advanced management techniques as identified through the National Response Plan and NIMS. Both large and small jurisdictions may find themselves in a situation where the use of advanced **Incident Command Management** would need to be used.

Now you should have a better understanding of incident typing and how the system can be used to determine the complexity of an event. We

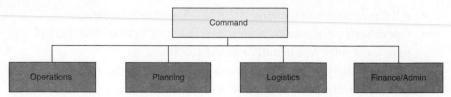

Figure 2-1   *ICS Command Staff and General Staff.*

will now expand upon basic ICS skills to address the management of major or complex events. To start with, all Command and General Staff positions will be filled. Figure 2-1 will identify those positions.

In Figure 2-1, the General Staff positions are listed below the Command. According to basic ICS training, the Incident Commander can create a Command Staff, which would consist of a Safety Officer, Public Information Officer, and Liaison Officer. During the Apex event, the Command Staff consisted of a Unified Command Team and staff positions that included the Safety Officer, a Public Information Team, and a Liaison Officer. These positions report directly to the Incident Commander or Unified Command and would appear in Figure 2-1 between Command and the General Staff.

Span of Control

Before we go any further, we need to discuss **Span of Control** as it relates to the major/complex incident. Optimal span of control is five. However, when managing a major/complex incident, span of control is expanded by applying recognized incident command practices to effectively utilize resources, but at the same time not exceed management capabilities. An example of this would be as follows: The Operations **Section** could have as many as five **Branches.** Each Branch could then have five **Divisions** or **Groups.** Each Division or Group could then have five **Task Forces** or **Strike Teams,** and finally each Task Force or Strike Team could have five **single resources.** This example shows how that many resources can be managed effectively with the Incident Commander never exceeding their allotted span of control. Before we continue, lets review the terms listed previously in this section.

- Division—the organizational level responsible for operations within a defined geographic area.
- Group—established to divide the incident into functional areas of operation.

- Branch—organizational level having functional, geographical, or jurisdictional responsibility for major parts of the incident operations.
- Task Force—group of resources with common communications and a leader that may be preestablished and sent to an incident; or they can be formed during the event.
- Strike Team—specific combination of similar kind and type of resources that have common communications and a leader.
- Single Resources—an individual piece of equipment and its assigned personnel or an established crew or team of individuals with an identified supervisor.

These are basic terms, but you need to understand them as the foundation or building blocks for the management organization of a major/complex incident.

General Staff and Support Positions

Under the General Staff positions, all support units will be filled.

Planning Section Organization

Under the Planning Section, there are four unit positions that can be filled. Figure 2-2 identifies how these units will be organized and who they report to. It is also important to understand their responsibilities.

- The Resource Unit is responsible for maintaining the status of all assigned resources, both primary and support, at an incident. The Resource Unit Leader can achieve this by overseeing the check-in of all assigned resources, by maintaining the status of all resources, and by the maintenance of a master list of all on-scene resources.
- The Situation Unit is responsible for the collection, processing, and organization of all incident information to include incident growth, maps of the event, and pertinent information surrounding the event.
- The Demobilization Unit will be responsible for developing the Incident Demobilization Plan. In the Demobilization Plan, the Demobilization Unit will monitor ongoing resource needs, identify surplus resources, develop the incident checkout procedures, and determine support needs for demobilization.
- The Documentation Unit is responsible for the maintenance of all incident files to include the accurate recording of all incident reports and forms. All Unit Leaders will maintain an ICS 214 form.

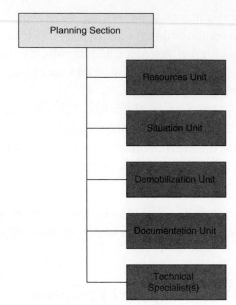

Figure 2-2   *Planning Section organization.*

- Technical Specialists may be required for certain major events where expertise on a specific type of event is required.

Logistics Section Organization

There are six units that may be assigned under the Logistics Section. You will notice that under the organizational format in Figure 2-3, the Logistics Section has been broken down into two Branches: the Service Branch and the Support Branch. Under the Service Branch there are three unit designations. The Branch positions as identified in the figure are not required but simply one method to expand the Logistics Section. When used, the Branch positions are responsible for the following.

Service Branch Director   Under the supervision of the Logistics Section Chief, the Service Branch Director is responsible for the management of all service activities. Under each Branch Director, there are three units assigned. The Service Branch Director's responsibilities are:

- Communications Unit—responsible for developing plans for the effective use of incident communications equipment and facilities. The Communications Unit Leader will maintain an ICS 214 and prepare the ICS 205.

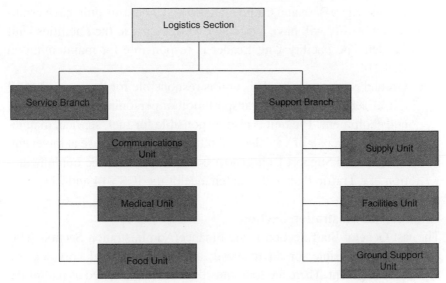

Figure 2-3 *Logistics Section: Two branch organizational structure.*

- Medical Unit—responsible for the development of a medical plan, obtaining medical aid and transportation for injured and ill personnel at the event, the establishment of responder rehabilitation, and maintenance of applicable reports. The Medical Unit Leader will maintain an ICS 214 and prepare the ICS 206.
- Food Unit—responsible for supplying the food needs for the entire incident to include any remote sites such as camps or staging areas. The Food Unit Leader is responsible for maintaining an ICS 214.

Support Branch Director   Under the direction of the Logistics Section Chief, the Support Branch Director is responsible for the development and implementation of the logistic plans in support of the Incident Action Plan. The three units assigned under the Support Branch Director are as follows:

- Supply Unit—responsible for ordering personnel, equipment, and supplies as well as receiving those supplies. The Supply Unit must be prepared to store and maintain supplies, keep an inventory of available supplies, and servicing equipment. The Supply Unit Leader will maintain an ICS 214.
- Facilities Unit—responsible for the layout and activation of incident facilities, such as a base, camp, and the Incident Command Post. This unit provides sleeping and sanitation facilities for incident personnel

and manages Base and Camp operations. Under this unit, each established facility will have a manager who reports to the Facilities Unit Leader. The Facility Unit Leader is responsible for maintaining an ICS 214.

- Ground Support Unit—This unit is responsible for the support of all out-of-service resources, transportation of personnel, supplies, food, and equipment. The unit is also responsible for fuel, service, maintenance, and repair of vehicles and other equipment in use at the event. The Ground Support Unit Leader is responsible for the implementation of a Traffic Plan and maintenance of the ICS 214 and 218.

Finance/Administration Section

The last General Staff Section is the Finance/Administration Section. This section is responsible for all financial, administrative, and cost analysis aspects of the event. There are four units that may be assigned as part of the expansion of this section as shown in Figure 2-4. Their responsibilities are as follows:

- Procurement Unit—This unit is responsible for administering all financial matters pertaining to vendor contracts, leases, and other financial agreements as they relate to the specific event. The Procurement Unit Leader is responsible for maintaining an ICS 214.
- Time Unit—This unit is responsible for equipment and personnel time recording and the management of the commissary operations. The Time Unit Leader will maintain an ICS 214.
- Compensation and Claims Unit—This unit is responsible for the overall management and direction of all administrative matters as they pertain

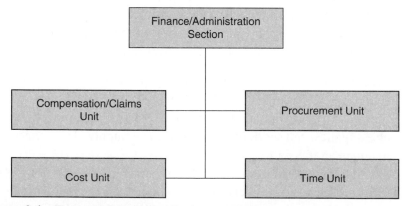

Figure 2-4   *Finance Administration Section organization.*

to compensation for injury and other claims-related activities other than injuries related to the event. The Compensation/Claims Unit Leader will review the Medical Plan and maintain an ICS 214.

- Cost Unit—The Cost Unit is responsible for collecting all cost data, performing cost analysis, and providing cost estimates for the event. The Cost Unit Leader will also make recommendations on how to achieve cost savings. The Cost Unit Leader will record all cost data and maintain an ICS 214.

Information/Intelligence Function

We have now addressed how to expand upon the basic ICS organizational structure. We have looked at how you can expand General Staff positions and use Branches to manage the many additional responsibilities that a major/complex incident may create. There is one new position that needs to be defined. This position is an additional responsibility role that can be added to several positions within your organization depending on the needs of the event. This is the **Information/Intelligence Function.** This position is responsible for the gathering of information and intelligence related to the incident. This may include, but is not limited to, national security matters, operational information, weather data, toxicity levels, utilities data, and structural design information. This function can be placed under the Command Staff, as a Branch under the Operations Section, as a Unit under the Planning Section, or as a separate Section under the General Staff. It will be the decision of the Incident Commander to identify the need for this function and its placement within the organization. Generally, the Information/Intelligence Function is placed under the Planning Section as a separate unit as shown in Figure 2-5.

Characteristics of the Major/Complex Incident

Now let's look at some specific characteristics of the major/complex incident and how to manage such an event. The major/complex incident will be resource-dependent, thus requiring large numbers of tactical and support resources. These resources will have to be ordered, tracked, and managed. The event will extend into several operational periods and require written Incident Action Plans. Depending on the duration of the event, Command will probably be transferred several times. The management of

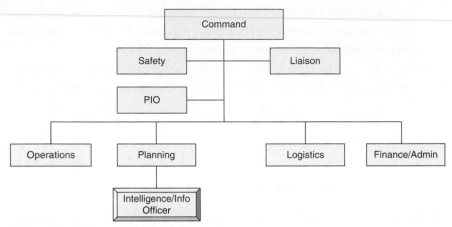

Figure 2-5   *Information/Intelligence function.*

such an event will require the use of ICS, but will have to be expanded upon so as to address the following items:

- Administrative and jurisdictional responsibility
- Geographical area involved
- Span of Control
- The requirement of specialty resources
- The potential for growth of the management organization

Tailoring the ICS Structure

The ICS structure will adapt to meet the needs of most large incidents but may have to be tailored to address specific management needs of major/complex incidents. Next are four possible options to the expansion and tailoring of this management tool. Each of the four options will be discussed in detail.

Incident Complex

An **Incident Complex** is two or more individual incidents located in the same general proximity that are assigned to a single Incident Commander or Unified Command to facilitate the management. This option is used when there are many separate incidents occurring close together, when during one incident other small events occur in close proximity, or when management can be facilitated by an incident complex. Several things must be considered when developing an Incident Complex. The incidents must be close enough together for the same Incident Management Team to manage them. A single

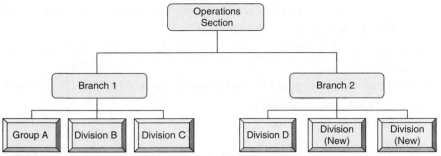

Figure 2-6 *Two Branch Organization.*

Incident Management Team can handle many of the other functions such as Planning, Logistics, and Finance/Administration. Basically, by using this option to manage several like events in close proximity to one another, you will consolidate required management responsibilities under one Incident Management Team. Figure 2-6 is an example of the Incident Complex Structure.

An Incident Complex is generally managed under a Unified Command with separate Branches under the Operations Section Chief. This allows for the establishment of Groups or Divisions if necessary. By using the Branch designation for each separate event, already established Divisions or Groups from each event can remain in place.

Division of a Single Event

At times, a single event becomes too large to manage and thus may require dividing it into two or more single events. A single incident may be divided when one or more of the following problems occur:

- The incident spreads into another jurisdiction and Unified Command is not feasible.
- Due to terrain and access to the incident, management from one location is not feasible.
- The objectives of the incident are naturally separated into two operations.
- The Planning and/or Logistics Section can no longer adequately provide support services.
- The Operations Section can no longer manage the number of resources required without exceeding span of control.

If one of the previous problems does arise, requiring the division of the single incident, follow these steps:

1. Determine how you will divide the incident based on the type of problem identified.

2. Assign Incident Commanders along with Command and General Staff to each new incident.
3. Identify facilities and a suitable location for each new Incident Command Post.
4. Designate the appropriate time to separate the incidents and define each new incident with a unique name.
5. Coordinate planning strategies and the use of critical resources for at least the next operational period.
6. You may also want to consider the use of Area Command (to be discussed later in this chapter).

Expand the Planning Capability

Another option for tailoring the ICS structure is to expand the Planning Section. With this option you can implement **Branch Tactical Planning.** This option is implemented when no single set of objectives is pertinent to the entire incident or when special expertise in a technical field is needed for planning. You may also find it necessary to implement Branch Tactical Planning if an Incident Action Plan cannot be prepared in a timely manner and be distributed to all operational personnel. In the use of branch planning, the Planning Section will develop the following:

- General incident objectives
- Strategy for the Branch in the next operational period
- Branch resource summary for the next operational period
- Weather and safety information
- Changes to logistical support
- Personnel to support planning

By providing this information, the individual Branches can then perform detailed action planning. The Planning Section will ensure that necessary coordination between branches takes place. Another way to expand on the planning concept is to perform advanced planning. Advanced planning can be achieved by assigning a Deputy Planning Section Chief to manage and oversee advanced planning. Advanced planning should project ahead 36 to 72 hours and consider the following:

- Overall goals and incident objectives
- Adequacy of Incident Action Plans
- Future resource needs verses availability
- Environmental considerations
- Long-term needs as they relate to the event and recovery

The goal of advanced planning is to provide the Planning Section Chief and Incident Commander with a range of alternatives for consideration beyond the next operational period.

Adding an Additional Operations or Logistics Section

The need to add an Operations or Logistics Section is a rare occurrence, but it is offered as an option to Command Staff when necessary. The option of adding an Operations Section can be used if the ongoing operation will require the existing Operations Section to expand beyond span of control guidelines. The other reason for this consideration may be geography where tactical operational resources are separated to a point that a single Operations Section Chief cannot effectively manage allocated resources. As rare as it may be, this was the situation addressed in the Environmental Quality (EQ) Fire. With operational resources separated geographically into high hazard environments, Command Staff decided to use the option of creating two Operation Sections. Command Staff may find it necessary to add a Deputy Incident Commander of Operations to aid in management. Just as rare is the addition of a second Logistics Section, but it may be required if the geography of the event makes it difficult for the incident base to support logistical needs of the incident.

Area Command

When your event expands beyond the capabilities of one of the aforementioned expansion options, you should consider using **Area Command** as a method to oversee the management of your event. Area Command is used to manage multiple incidents handled by an Incident Command System or a single large event that has multiple Incident Management Teams already assigned. Area Command can be very useful when there is a number of events occurring in the same area and of a similar type. Some examples may be two hazardous materials events or two wildfires. These are two good examples of similar events that generally use the same types of resources. What happens in the development of Area Command? The Incident Commander performs primary tactical level, on-scene incident command functions. The Incident Commander is located at an Incident Command Post at the incident scene. The **Area Commander** oversees the management of multiple incidents. Area Command may be unified and works directly with Incident Commanders. The Emergency Operations Center acts as the coordination entity coordinating information and resources to support local incident management activities. By using Area Command you will help coordinate interagency activities. Area Command is very useful in making sure that resources are used efficiently between multiple events, each requiring similar resources. The use of Area Command will ensure that all agency policies, priorities, constraints, and guidance are communicated to the Incident Commanders and that they are implemented consistently across incidents.

When Area Command is established, Incident Commanders will report to the Area Command. The Area Commander is accountable to the agency or jurisdictional executive. If more than one jurisdiction is involved, then the Area Command may need to be implemented as a **Unified Area Command.** Incident Commanders operating under the designated Area Command are responsible to and should be considered as part of the overall Area Command organization. Incident Commanders should be provided with clear authority and all delegations of authority should be in writing. The Area Command will be responsible for:

- Setting overall event objectives.
- Ensuring that incident objectives are met and do not conflict with agency policy.
- Establish incident-related priorities.
- Allocate/reallocate critical resources based on incident priorities.
- Ensure that Incident Management Teams are qualified.
- Coordinate demobilization of resources.
- Coordinate with the Agency Administrator, EOC, and other Multi-Agency Coordination Centers, as well as the media.

All Area Command authorities should be delegated in writing. The following should be considered as direction for establishing Area Command:

- When several activities are in close proximity.
- When critical life-saving or property values are at risk due to incidents.
- When incidents are projected to continue into a second or additional operational period.
- When each incident is using similar, limited resources.
- When Incident Commanders encounter difficulties with interagency or interincident resource allocation and coordination.

Figure 2-7 shows an example of the organization of a Unified Area Command.

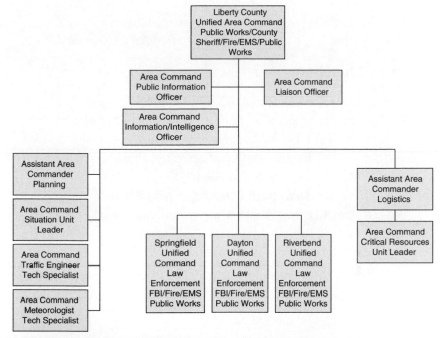

Figure 2-7 *Example of a Unified Area Command Organization.*

Chapter Summary

Chapter 2 elaborates on ICS, taking into consideration that the concept of basic Incident Command is a skill that most have already learned and used in their everyday experiences. The chapter delved into the not-so-routine events and how the basic ICS could be enhanced and built upon to deal with bigger events that may exceed the capabilities of local responders. Complex Incident Management and Area Command have been explained and examples given of how each role within the command organization can be expanded upon.

Key Terms

Major/Complex Incident	Group
Incident of National Significance	Task Force
Incident Command Management	Strike Team
Span of Control	Single Resource
Section	Information/Intelligence
Branch	Function
Division	Incident Complex

Branch Tactical Planning   Area Commander
Area Command   Unified Area Command

Review Questions

1. Identify three characteristics of a Major/Complex Incident.
2. What is the definition of an Incident of National Significance?
3. Under the NIMS Typing Criteria, emergency events are divided into five types based on several specific characteristics. What would characterize a Type 2 incident?
4. What is the Service Branch Director responsible for?
5. What is Area Command, and when would you use it?

Chapter 3

Local Area Hazard Assessments— The First Step to Pre-Event Planning

> *"A good plan, violently executed now, is better than a perfect plan next week."*
>
> *General George S. Patton Jr.*

Learning Objectives

After reading this chapter, you should be able to

- Define pre-event planning
- Identify the local resources that can help with the hazard assessment within your community
- Identify locations suitable for emergency shelters
- Identify ways to educate and involve your community in the area of emergency preparedness

After hearing about the explosion and chemical fire in Apex in October 2006, groups in the emergency services arena, including fire, police, EMS, and emergency managers, all came up with the same question: How did you do it?—referring to the effective response, mitigation, and stabilization of the event. At first, I had to think about the answer. But looking through the events of that weekend, the answer was clear. It could be summed up in three words: planning, training, and implementation. This chapter will explain how to assess your jurisdiction and identify potential problems.

Pre-Event or Pre-Incident?

According to the USFA's official definition, an event is a future activity that will include the activation of an ICS organization. An incident, however, is defined as an unexpected occurrence that requires immediate response actions through an ICS organization. This definition is specific to the USFA as it relates to the use of the Incident Command System.

Whether you are planning for an event or an incident, you will need to consider the following:

- Type of event
- Location, size, expected duration, history, and potential to be able to project objectives
- Number of agencies involved
- Whether the event/incident will be single or multijurisdictional
- Command Staff needs
- Kinds, types, and numbers of resources
- Facilities needed to include staging areas
- Communications and financial considerations

In addition to the items listed, there are other factors to consider for an emergency incident. The emergency incident is time critical. The emergency incident is also an unstable situation that is ever changing and has the potential to expand rapidly. For these reasons, the emergency incident requires more immediate actions to ensure effective incident management and response to gain control over the situation. For the purposes of this text, we will address and reference incident planning, but you should understand that the material can be applied to both an incident and an event depending on the type of situation you are planning for.

Pre-Incident Planning

The key factor affecting the outcome in Apex and what should take place in every jurisdiction is **pre-incident planning.** Pre-incident planning is the act of identifying what potential problems you as a responder may be faced with in your jurisdiction and then developing a plan that is "all hazard" in nature. During the planning phase, you should visit occupancies within your jurisdiction and identify what hazards they hold and what threats they pose. Your local planning department or Fire Marshal's office may be helpful with this. Both of these offices generally have business occupancies on file. You can also identify such occupancies from your Local Emergency Planning Committee. However, you must realize that business occupancies are not the only areas that your plan should identify. Other areas would include flood-prone areas, large thoroughfares with high vehicle traffic, rail lines, airports, high tourist areas with occupancies holding large numbers of seasonal guests, and any types of weather events that frequent your jurisdiction.

Resource Sufficiency Assessment

Once you identify the areas that your jurisdiction will respond to, you can perform a **resource sufficiency assessment.** This is where you look at your response capabilities compared to what each area of concern will require for an effective response.

Here is a simple example of resource sufficiency assessment. A building in your jurisdiction requires a 2,000 gallon-per-minute fire flow. However, when you look at your available resources, you can only produce 1,500 gallons per minute. Based on this assessment, you will need an additional fire apparatus from a neighboring department to produce the required fire flow. Identifying this type of problem before a potential hazard is the key to true preparedness.

Remember, when you perform your assessment of available resources, leave no rock unturned. The plan should not just look at the fire department or other emergency services agencies. Consider all the resources available to you. For example, Apex is a full-service town, meaning services include fire, police, EMS, electric utilities, water and sewer, environmental monitoring, planning, IT services, and code enforcement. When we performed our resource assessment, we looked at how each of these services could be applied in response to certain events. During the October explosion, we applied all these services to enable us to respond to the event. (Further details on how these services were employed are discussed later in this chapter.)

Once you have identified all potential resources, you need to determine if your resources will be sufficient. Understand that you may not have every resource that you need, so you must now determine what types of assistance you will need. You should then look at your neighboring jurisdictions and see if they can provide any of the resources you may be lacking. In doing so, you can enter into mutual aid agreements with these agencies that will bring the necessary resources to you when you need it, and vice versa. By having certain necessary resources pre-identified, you alleviate the chance of delays in receiving these resources.

Identifying Shelter Sites

Another issue that your Incident Plan needs to address is where you can house people. People can quickly become part of any emergency that you respond to. For example, a severe weather event or chemical emergency can easily displace a large portion of your jurisdiction in an instant. As

Figure 3-1   *One of three emergency shelters opened during the Apex incident. A total of three schools were used to house evacuees. Courtesy of Mike Legeros, photographer.*

part of your plan assessment, you must identify available building space for these people. The allotted space may be a local community building, church facility, or other town building. When identifying space for evacuations, you must consider the issue of backup power capabilities and space for overnight stay. The space should have facilities for showering and restrooms, in addition to cooking and food preparation areas. A place that usually offers all these amenities is a local school (see Figure 3-1). Developing a working relationship with your local school board will aid you in the acquisition of these spaces when needed. You will want to obtain a contact list of school board members so that you can contact them if needed. Remember that when using school facilities, you will have to consider the effects of this acquisition. Students will not be able to use the school if you are. This needs to be worked out prior to the event and necessitated use.

Notifying the Community

You have identified the potential problems and the needed resources. You have researched whether or not your current resources are sufficient. If not, you have learned where to obtain them, and most importantly, what to do with the community population that the event may affect. But there is one other thing: How do you effectively notify your community? There are many proven ways to do this, yet it seems that this is where most large-scale events start falling apart. The key to successful community notification and evacuation is, once again, pre-event planning. Hold events such as a "Community Awareness Day," where citizens come together with emergency response agencies. You can distribute such items as a Family Emergency Plan, Pet Action Plan, routing instructions for emergency shelter locations, and evacuation routes. For examples of these items, see Appendices B, C, D, or the accompanying NIMS Resource CD. By bringing your community together, everyone will be more prepared in the event the plan ever has to be enacted. Events such as this make the citizens an integral part of the emergency response effort. No, the general public will not be part of your actual emergency response staff, but they will be a part of the response plan by knowing how to get out of harm's way to a safe haven. This takes one thing to worry about off your plate.

Modes of Communication

During an actual incident, you may choose to simply use what is already available to you in your jurisdiction, such as radio or media. You may already have public notification systems such as a fire department siren or other emergency management siren systems in your jurisdiction. If these are not options, there are some other means that will work. During the EQ Fire, the townwide 911 system was contaminated due to vapor cloud. The North Carolina Canine Emergency Response Team (CERT) Communication Unit was utilized to provide an emergency mobile communications center. See Figure 3-2.

Systematic Route Alerting

One effective alternative to radios and siren systems is **systematic route alerting.** This is a system where you identify routes within your jurisdiction and map them out. Then, local emergency responders run the route in a quick and effective manner, using loud speakers on emergency vehicles in conjunction with vehicle sirens. At strategic points along the route, the siren will be sounded followed by a verbal message that has been predetermined. An example of such a message would be as follows: "An emergency has occurred that requires immediate action to insure your safety. Please tune to your local radio or television station for further information and/or report to your identified emergency shelter for further instructions." Of course, for

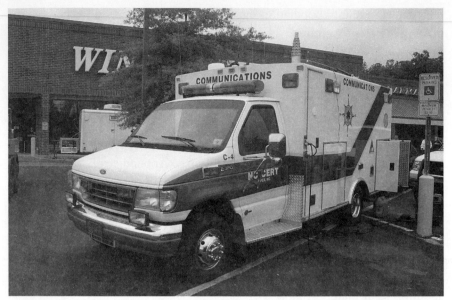

Figure 3-2   *North Carolina CERT Communications Unit in operation at the Apex Incident. Courtesy of Mike Legeros, photographer.*

this to be effective you will have to have already worked out an arrangement for local media to assist you with public emergency messages.

Implementation in Apex

In the previous pages, we have discussed the concept of pre-incident planning and how it can help you to identify potential hazards in your response jurisdictions. This is exactly what was done preceding the chemical fire in Apex. Let's look at some of the pre-incident actions that were taken as part of the Apex planning process.

An Icy Warning

Of course, the initial incident that sparked reform was an ice storm in December 2002. During the storm, the town experienced a large transformer fire and, in turn, lost power to approximately half of the city for nine hours. The fire department was able to extinguish the fire, but not before the substation had experienced extensive damage. While electrical crews worked diligently to repair the fire damage and get the transformer back online, it became evident that many other issues needed to be addressed. There was no plan for this type of incident, nor was there a predetermined location for an Emergency Operations Center or a citizen shelter. Fortunately, the fire department had just opened a new fire station with a large training room. This room was used to serve as an emergency operations center, and the local community center was opened to serve as an emergency shelter for two nursing homes and some 45 evacuees who had no supplemental heating source during the power outage. This incident worked out well with power being restored within 10 hours, and no fatalities or injuries. Apex was able to successfully get through this incident, but we realized that there were many areas in our existing plan that needed to be addressed for future incidents.

Identifying Potential Threats

Realizing that Apex's plan needed updating, we started to work toward identifying the area's target hazards and rewriting the plan to address these identified areas of concern. We realized that the plan needed to be "all hazard" in nature and should address every type of potential incident that may occur in Apex. The plan needed to identify how citizens would be notified during an emergency incident as well as how evacuations would take place. Sheltering sites needed to be identified, and staff had to receive training in how to manage emergency shelter sites. Contact information and an effective organizational structure needed to be updated and maintained. Within the plan, there needed to be an identified site for the Emergency Operations Center. The Town Manager agreed that the fire department training facility would serve well as an Emergency Operations Center. With the manager's approval, staff began upgrading this facility with additional network lines, dedicated phone lines that would only be used during an emergency incident, and an emergency phone connecting directly to the 911 center and radio communications. A weather monitoring station was also added to allow for real-time weather tracking. The fire department then started the target hazard identification process. They identified many hazard concerns around Apex that needed to be part of the Emergency Plan. Following are some of those hazard concerns:

- Harris Nuclear Facility—this facility was identified because it is a nuclear power facility, and Apex was in the Emergency Planning Zone.
- Dixie Pipeline—this facility is a large commercial propane transport and storage site with supply pipelines running through the town.
- Motiva—this facility stores fuel products and manufactures butane.
- CSX Railroad—the railroad runs directly through the entire Apex Fire District. Daily freight and passenger trains come through town.
- Environmental Quality—A Type I Hazardous Waste Handling Facility located in the industrial corridor of Apex.

- Downtown Business District—this was identified because of the large fire potential due to the age of many of the businesses located in this area. The downtown business district is the oldest part of Apex, dating back to the late 1800s.
- Highway Hazards—this was identified in reference to the three major thoroughfares that have the potential for commercial vehicle accidents including hazardous materials incidents.
- Weather Events—Apex is only 128 miles from the coast and during the summer has tropical storms and hurricanes. The area also has winter weather events that bring their own problems.

This is not the entire list, but as you can see, many areas were identified during the hazard assessment. The identification was part of the "all-hazard" process for developing a plan that would prepare responders to address any type of emergency situation they were faced with.

Rewriting the Plan

After the hazard identification was completed, the plan was rewritten to address responder responsibilities, departmental objectives, pre-incident preparation, response concerns during the incident, post-incident actions, and public/community concerns for each related incident type. Addressing the aforementioned areas allowed the town to create a plan that followed other state and federal guidelines, which made the plan more uniform. It also addressed each department's responsibilities in these areas of response, which were not included in the original plan. All departments within the town's organizations were trained in how to use the plan. As part of the training, the police and fire departments held simulated emergency drills and tabletop exercises in an effort to insure the plan's effectiveness and that our personnel were well versed in the use of the plan. In October 2005, the town received a grant to bring the plan into compliance with new NIMS guidelines. This was completed in May 2006 and was distributed to all departments, which were required to go through educational training on the use of the plan. All personnel were required to receive applicable incident command training as well, which is used as the recommended management model for emergency incidents in Apex.

The training was completed in July 2006, only two months before the EQ Fire and explosion. By having a plan that addressed the many actions that had to be employed during the EQ chemical fire incident, and having responding personnel that were trained in how to use the plan, this incident

Figure 3-3   *Resources in the Staging Area. Courtesy of Lee Wilson, photographer.*

ended successfully. Of course, the plan was only part of the process, but it was the beginning to a successful end. The fact that Apex had required responding personnel to be trained in not only how to use the plan but also how to manage the resources needed to mitigate this type of incident was instrumental to the success of this response. See Figure 3-3, which shows the Staging Area located at the fourth Command Post. From the Staging Area, responders could see explosions over the tree lines. The photo offers an exhibit of the number of resources that responded and identifies one of the many challenges that personnel were faced with and had to manage effectively.

Chapter Summary

Chapter 3 defined pre-incident planning and provided some examples of facilities that should be reviewed and situations to be planned for before an incident occurs. Pre-incident planning can be performed for both man-made facilities as well as natural events, such as tornadoes, floods, and hurricanes. Pre-incident planning played an instrumental role in the successful outcome of the EQ Fire. We also discussed how

Apex had implemented pre-incident planning in its jurisdiction long before this incident ever occurred. You should understand that this process started right after the ice storm event and was a work in progress until 2006. This is an important part of any plan—leaving it open for continued improvement and training as the area changes through growth or continued community development.

Key Terms

Pre-Incident Planning   Systematic Route Alerting
Resource Sufficiency Assessment

Review Questions

1. Define pre-incident planning.
2. Where can you find information that will help you to identify potential hazardous occupancies in your response area?
3. Identify three things that you would look for when identifying appropriate emergency shelter sites.
4. Based on the information provided in this chapter, identify at least three businesses within your response district that would require pre-incident planning.
5. Identify some key equipment needs that should be considered when creating an Emergency Operations Center facility.

Chapter 4

Pre-Incident Planning and the Emergency Operations Plan

"Plans are only good intentions unless they immediately degenerate into hard work."

Peter Drucker

Learning Objectives

After reading this chapter, you should be able to

- Identify how pre-incident planning works in conjunction with the definition and creation of your Emergency Operations Plan
- Define the differences between Standard Operating Guidelines and an Emergency Operations Plan
- Identify sections within your Emergency Operations Plan as they relate to your local area hazard assessment
- Identify, develop, and understand your Emergency Operations Plan
- Identify the concepts of "flow of requests" and event activation

Before you can successfully respond to any type of situation, you must first have a plan. Look at World War II for example. The Japanese military developed a foolproof plan to attack the United States Navy at Pearl Harbor. Then, in 1944, the Allied Alliance developed a plan for D-day and successfully attacked the German Army. On a smaller, local scale, let's look at holiday events. Each year the Macy's Thanksgiving Day Parade and the New Year's celebration in Times Square are carried out without a hitch. The events are managed and the crowds taken care of. What do all four of these examples have in common? Every detail and everything that could potentially go wrong has been addressed and planned for in advance.

By using the aforementioned familiar events, we can all agree that planning is a key factor in successful outcomes. Do you think any one of the four events would have turned out successfully if a plan had not been intricately developed for the desired outcome? I don't think so. But, why did I choose to use nonfire events as examples of effective planning? The reason is that planning can be used for anything from building a house to having a downtown parade to responding to a large-scale emergency.

Standard Operating Guidelines

In the fire service, we plan daily and respond to every call under some type of a plan. The plan may not be directly related to your particular call, but even departmental operating guidelines are a type of plan. They are written with an objective, a desired outcome, the necessary strategy to implement, and the tactics that will achieve the desired outcome. Do we have them for every type of call we respond to? No, but we generally write **Standard Operating Guidelines (SOGs)** for a wide range of call types such as EMS calls, structure fires, and so forth. So we use this type of planning to deal with our daily "bread and butter" calls, but the Emergency Operations Plan (EOP) will be the plan that is used to address the not-so-common call. In this case, you or your department may have to respond to and manage an event requiring numerous resources and a methodology to manage those resources, thus creating a planned response that will identify both the response strategy and the tactics to be used to bring the event to a successful end. Will it be without fault? Probably not. Will it be without problems? Probably not. But with a plan that addresses these issues, you can overcome such obstacles and move on. There will always be problems, and you as the Incident Commander or Manager of the event will be looked upon to resolve each and every thing that occurs. Remember it is the Incident Commander who is in charge until the roles and responsibilities are designated to the Command and General Staff.

Let's look at what we must do to develop a plan and certain elements that have to be addressed when writing that plan. See Figure 4-1 for an example of an aerial photo that may be used in your plan to identify a specific hazard concern. In this photo, you can see the EQ facility surrounded by other commercial businesses and a bordering residential neighborhood.

Perform a Hazard Survey

The first item on your agenda should be to perform a **hazard survey** of your response area. This hazard survey will identify what types of potential emergencies your department may have to respond to. For example, if you have an airport in your area, your survey should address

Figure 4-1 *Aerial photo of the Apex industrial corridor. Reprinted by permission of Google Earth Mapping Service.*

response to a downed aircraft. If there is a railroad close by, the survey should address how you would respond to an incident involving a train. This could include a derailment, possible chemical release, or passenger train accident with numerous casualties. You may have a high-traffic roadway that presents such problems as a transportation accident or motor vehicle incident with multiple injuries. Any of these potential hazard zones could be present in your area and thus should be addressed in the survey.

Hazard Survey Resources

A local resource that can be used to assist agencies with their hazard survey is the **Local Emergency Planning Committee (LEPC).** Every city and/or county, depending on governmental design, should have an LEPC as required by the Code of Federal Regulations 1910.120. Under this standard, local government should develop an LEPC that is generally comprised of emergency officials, local business leaders, and citizens who have input on identified hazards within their jurisdiction. The LEPC will usually have a list of potential high-hazard facilities already identified and should be able to share this information with you. Some other agencies that may assist you in your research are the local building inspections office or planning office where new construction plans are filed. These construction plans will identify the type of business and whether there is a hazard to be considered.

As you identify potential hazards that the plan should address, don't forget to look at the surrounding area. For example, if you identify a manufacturing facility as having a high-hazard occupancy, you should also look to see if there are any neighboring properties such as schools, residential subdivisions, or other businesses to be evacuated. You should consider this as part of your planning process.

Once you have identified the target hazards that your plan should address, you will need to look at your available resources both interdepartmentally and within your jurisdiction. Based on your hazard survey, do you have the necessary resources to effectively respond to the identified incident, or will you need to ask for assistance from other jurisdictions? Whether you are a volunteer department or part of a local government structure, you will need to identify all the resources that are available to you. Here are some questions to consider:

- If you have a public works department, do they have heavy excavation equipment and dump trucks that could be used for containment operations during a chemical release?
- Does the local police department have the staff available to assist with a forced evacuation and the maintenance of an exclusionary zone?
- Who provides power and telephone services, and can you get in touch with them if needed?
- Does your jurisdiction have a water and sewer department? If so, will they have water quality technicians and environmental professionals to help you with decisions regarding runoff?
- Who handles emergency management in your jurisdiction? This individual or group of individuals will be an important resource because they generally have access to resources outside that of local jurisdictions and will play an important role in assisting your Logistics Team in finding the resources needed.

Clearly, developing a pre-incident relationship with your local Emergency Management Office and other agencies can pay huge dividends when facing an incident.

Developing Your All Hazard Plan

Now that you have identified the possible hazards that your plan should address and developed a resource inventory for your local jurisdiction, you need to record this information as the first step to developing your **All Hazard Plan.** Within the plan, you will identify an individual section for each identified hazard. Here is an example of this:

- Section 1—Hazardous Materials Release
  A. Train Derailment
  B. Highway Incident
  C. Fixed Facility Incident
- Section 2—Weather Emergencies
  A. Tornadoes
  B. Hurricanes
  C. Floods
  D. Winter Storms
- Section 3—Other Emergencies
  A. Civil Disorders
  B. Terrorism
  C. Community Health Situations

This is just an example of how you may organize your plan to address the multitude of potential hazards and threats that your assessment identifies. Under each of the headings, you will identify high-hazard areas and what resources would be called upon to successfully respond to and mitigate the situation. See Figure 4-2 for an excerpt from the Town of Apex Emergency Operations Plan. It shows support functions and provides basic information and concepts for coping with potential hazardous material incidents, chemical, biological, radiological, nuclear, or explosive CBRNE (Chemical Biological Radiological Nuclear and Explosive) hazards within Apex. This document establishes a plan of action for coordination and support of emergency response operations, as required pursuant to the Superfund Amendments and Reauthorization Act of 1986 (SARA) Title III, also known as the Emergency Planning and Community Right-To-Know Act of 1986, Section 303(c). This support function is an integral part of the Town of Apex Emergency Operations Plan.

This brings us to the next section of the EOP, which each resource or agency will complete. Each agency or resource identified will need to create an objective-based list of responsibilities based on how they will respond

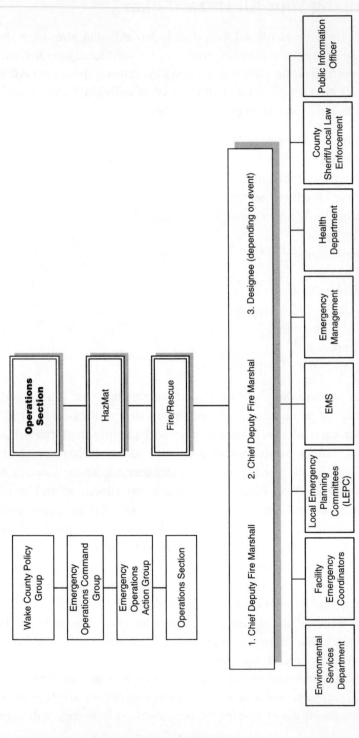

Figure 4-2 *Excerpt from the Town of Apex Emergency Operations Plan. Courtesy of the Town of Apex.*

to the threat. They will also need to include an organizational chart of their department, itemizing who is responsible for each task. This is important because it will help emergency managers/incident commanders develop an overall incident organizational chart and Incident Action Plan. Both of these items will be discussed in detail in following chapters.

Once you have your area hazards identified and what necessary resources you will need to respond to those identified hazard concerns, you will need to compile the data into a planning format. The plan should be written to be compliant with NIMS and Presidential Directive 5. At the end of this chapter are two Web sites that can assist you with finding information about NIMS and plan development. As you write your plan, remember it is supposed to serve as a guide, not a directive. It should guide department heads and personnel through the successful mitigation of an event, but it should not take away from their individual decision-making capabilities. Each situation will be different and thus will require an individualistic approach to the effective management and handling of the problem. The plan simply provides emergency managers with a tool that identifies possible strategies and tactics that they can use to bring the situation to a successful end.

Key Components to Your EOP Plan

As you develop your EOP plan, make sure you identify the following key components:

- List of target hazards with emergency contact information—this list will identify facilities you have performed pre-incident surveys on and have identified as a potential threat. You will list emergency contact information for each facility's representatives.
- A **threat analysis** that will identify natural and man-made hazards and what surrounding areas would potentially be affected.
- A resource list where you will identify what available resources your jurisdiction can provide in response to an emergency situation. You may want to also identify neighboring jurisdictions and any special response capabilities that they can provide.
- Organizational charts for each resource that are specific to each organization.
- Responsibility list that will identify who will take care of specific tasks. This needs to be practiced or at least discussed and understood. People will not function adequately in the position if the first time they do it is during an actual event.

- An **Evacuation Plan** that also identifies possible emergency shelter locations. The Evacuation Plan should include a methodology for the movement of citizens within your jurisdiction and the routes by which they would be moved. There should also be considerations listed for the need to "shelter in place" and how to accomplish this.
- A defined statement of authority. The City or County Manager, or another identified government official, generally provides this. This document will identify who has decision-making authority, which will reduce confusion during the actual incident.
- A defined declaration document. This would be a form document governing officials use to declare a State of Emergency. You can find a sample of this document in Appendix E.
- **Mutual Aid Agreements**—these are prearranged agreements between neighboring agencies that identify response capabilities, resources, and how they will respond when called on.
- Guide to how to exercise the EOP—this guide should identify objectives that need to be met when exercising the EOP. The guide should also list what each represented department should consider when training to use the EOP.

Functional Areas of Activity

There are some other considerations that you will want to think about when creating your plan. The first is to break your plan down into functional areas of activity. They are as follows:

- **Mitigation**—mitigation activities are designed to prevent or minimize the effects of a potential emergency situation. This can be accomplished through public education, community preparedness activities, public health directives, and code enforcement ordinances that are used daily. Mitigation efforts are largely derived from federal legislation. The Disaster Mitigation Act of 2000 (DMA 2000) identifies requirements for state and local mitigation planning. DMA 2000 provides funding for hazard mitigation planning. The mitigation area of your plan should identify preventive measures:
  1. Preventive measures that will reduce vulnerability of your community to identified hazards such as a weather event, flooding, or a chemical release.

2. Property protection measures that identify ways by which a community can protect existing structures or other vulnerable locations.

3. Natural resource protective measures that will center around protecting and preserving natural areas.

4. Identify structural projects designed to lessen the potential impact of a particular hazard by modifying environmental areas to reduce hazard impacts.

5. The last consideration is public information used to educate citizens, business owners, potential property owners, and visitors to the community of both the hazards associated with an area and the identified mitigation efforts in place to protect against those hazards.

- **Preparedness**—under this section, you will create activities and programs that will be used to train and prepare both your personnel as well as the community prior to an event. These activities and programs should be designed to support and enhance your emergency response. Some examples would be departmental training and education on the use of your plan, creating a tabletop of your community that could be used to simulate an emergency response or developing a volunteer citizen response group, such as Citizen Corps, to assist in the event of a large-scale event. The difference between preparedness and mitigation is that preparedness is the training required to be ready for an event, and mitigation is the actual effort to protect against or alleviate a problem. An example of preparedness would be identifying that a levee is getting ready to fail, and the mitigation would be taking the steps to protect against flooding and/or to remove the possible affected population and livestock.

- **Response**—response activities are actions based on determined strategies designed to address initial and immediate actions that must be taken during the onset of an emergency in an effort to gain control of the incident. Your response strategy should target methods that will reduce injury and both environmental and structural damage. Your response activities should also address methods to enhance and speed up recovery after an emergency. Some items addressed in this section would be command and control, emergency warnings, community evacuation, and mass health care.

- **Recovery**—your recovery activities should address **Continuity of Government** and **Continuity of Services.** You should address Continuity of Government in your plan by identifying how local government will continue to operate during and after an event. For example, during the Apex fire, all local government facilities were inaccessible due to possible chemical contamination. However, mobile facilities that included large tent structures were erected in a shopping center parking lot to provide work spaces for local government leaders to carry out necessary functions. Continuity of services should be addressed by identifying how basic services will continue after a catastrophic event has occurred. Citizens will expect certain services such as water, utilities, and transportation. Additional resources may have to be brought into a community to assist with such service provisions and to also provide expected emergency services such as fire, EMS, and law enforcement. You will need to identify how to perform damage assessment, who will perform such assessments, and how they will document their findings. You will also need to identify long-term recovery as well, such as neighborhood redevelopment, environmental impacts, and health concerns.

As part of your plan development, you will need to address how additional resources will be requested and received in support of your response and recovery efforts. You will need to identify contacts at the local, state, and federal level and ensure that staff personnel understand how to order these supporting resources. Generally, resources are initially received through local mutual aid from neighboring agencies, such as a police or fire department from a neighboring city. Once local resources are depleted, the local EOC can contact the state EOC for resource support. The state EOC will work with federal agencies to acquire additional resources and at this level can activate intrastate and interstate mutual aid through an Emergency Management Assistance Compact (EMAC), which will then make available federal funding, equipment, resources, and subject matter experts. See Figure 4-3 for an example of the flow of requests.

Response Phases

The **response phases** will identify at what stage of the event that certain actions should take place. Below is an example of how to phase your plan into action:

- **Increased Readiness**—Under this section, you will be monitoring a potential situation as it develops. This would be your "hurricane

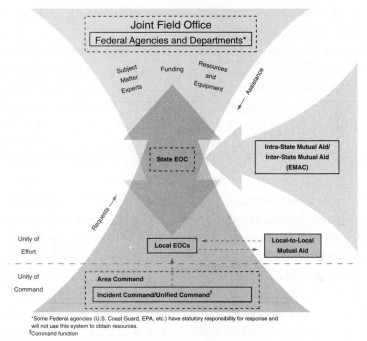

Figure 4-3 *Flow of Requests and Assistance During Large-Scale Incidents. Courtesy of the United States Fire Administration.*

watch." This phase may start up to 72 hours before the situation occurs and continue until the threat is present. In a fire or hazardous materials scenario, your personnel can be performing duties related to this phase daily through training and pre-incident surveys of your response jurisdiction. In preparation for a weather event, one may start monitoring the movement of a particular weather system to identify if the system is a threat to your community. As part of your monitoring activities, you will want to share information on the approaching storm or weather system with key personnel so that they can be aware of what is taking place. You may want to meet with staff personnel to review key components of your plan.

- **Pre-Impact**—this section relates more to a weather incident. This section is generally enacted once you have determined that impact is imminent. You would enact this section 24 hours before impact if possible. Under this section, you would cancel days off, recall all off-duty personnel, secure station grounds, check all equipment for a state of readiness to respond, prepare shelter locations, assign areas of responsibility based on the plan, and hold an initial staff meeting to determine readiness.

- **Impact Response**—this section defines actual response during an incident. Whether it is a weather event or man-made situation, this section is important because it will offer guidelines on how personnel should respond during the event. You should define when personnel will respond. You may be thinking, "we respond all of the time to every call for assistance." I would agree except that in certain situations it may be in your best interest to wait for a safer time to respond. For example, during the pre-impact phase, you have opened shelters and evacuated affected areas. Then, during the storm you receive a call for assistance, but winds are 75 miles per hour. You may want to hold your response until you can guarantee the safety of both responders and the community. Or if you are called to a hazardous materials event and you have sheltered certain areas in place, but then receive a call for assistance in the Hot Zone, then you will have to make a command decision about sending personnel into that affected area. These are considerations that have to be identified as part of your plan. You must have guidance for your personnel about when they can or cannot respond in these types of situations.
- **Sustained Emergency Phase**—under this section you will address operations post-event. How will Continuity of Government and Continuity of Services be reinitiated and maintained? Emphasis should be placed on helping injured and affected citizens, responding to the needs of displaced individuals, and addressing identified hazardous areas of your jurisdiction. Areas to be addressed in this phase would include, but are not limited to, communications, response, damage assessments, and environmental considerations.

Revisiting the ICS and EOC

So far we have discussed plan development, laying it out in a manner that will be functional and easily understood. The last two areas to discuss are the use of ICS and the establishment of an EOC. Under Presidential Directive 5, all agencies had to be compliant with applicable phases of NIMS by 2007. Part of this directive outlines that all events will operate under the ICS management system. Now let's look at what should be defined in your plan. Your plan should define how to use ICS and what positions to activate to manage the incident.

Levels of Activation in an EOC

Your EOC is the location used to coordinate information and resources to support local incident management activities. You will also have to outline in your plan when to activate your EOC and at what level of operation it will function. An EOC can be activated for both emergency and nonemergency situations. An EOC generally has four levels of activation:

- **Level I Activation**—This level is activated whenever your jurisdiction is notified of an event that may develop to threaten the safety of the general public (generally a monitoring status—not a full activation).
- **Level II Activation (incident-specific)**—This level is generally specific to one group and is activated in response to a hazard-specific incident that requires response from a specific group such as fire, law enforcement, or EMS. The Fire Chief, Police Chief, or EMS Director can activate the EOC at this level.
- **Level III Activation (limited EOC activation)**—This level is generally activated in response to a major incident and preempts full activation. Under this level, Command Staff will request initial personnel to be notified and respond to the EOC.
- **Level IV Activation (full activation)**—This level will build upon the foundation of the Level III Activation and incorporate government officials from the local jurisdiction as well as state and federal agencies.

Putting It All Together

Now that you have all the information needed to compile a plan, you will need to create your plan with respect to what hazards you have identified. Remember, you do not have to reinvent the wheel. There are likely many agencies that will be willing to assist you. Check with your local or state Emergency Management Coordinator to see if there is an existing plan that you can build upon. Another option may be to build your plan in conjunction with a neighboring jurisdiction. Because the requirement to have a plan in place and that the plan must be NIMS compliant is part of a Presidential Directive, there are grants available through state and federal agencies that will assist with plan development.

After your plan is complete, you must train all identified personnel on how to use the plan. All personnel should be well versed on their responsibilities and roles as they are defined within the plan. Training will provide agency personnel with the knowledge, skills, and abilities needed to use the plan during an emergency response. Training should include identifying individual tasks, team responsibilities, and situational needs. Your training should also include drills and exercises as well as critiques to identify lessons learned. The critique is important in identifying any problems with the plan and how to fix those problems before an actual incident occurs.

That is exactly how Apex had prepared. The Fire Chief initially rewrote the town's Emergency Operations Plan after an ice storm struck the city in 2002. In 2005, the Fire Chief received a grant that assisted Apex in creating a revised, compliant EOP that was completed in June 2006. The plan was tested on October 5, 2006, to the fullest extent and was determined to be successful.

The success was derived from three key components: an effective hazard survey, the creation of a response plan that defined how to respond to the identified threats, and training on how to use the plan. Below are some Web sites that you can visit to find more information on NIMS and EOP development.

- www.fema.gov/plan/index.shtm
- www.nimsonline.com

These Web sites will provide you with templates and sample documents that you can use in the development of your plan including an interagency mutual aid agreement and local government emergency operations plans. Some of the more frequently used documents can be found on the accompanying CD. There is also a sample EOP located in Appendix B.

Chapter Summary

Chapter 4 reviewed the difference between Standard Operating Guidelines and the Emergency Operations Plan. The development of an Emergency Operations Plan has been discussed in detail to include staffing of the EOC, levels of activation, and considerations for training of personnel in the use of the plan during an emergency event. The chapter also discussed what should be included in the Emergency Operations Plan, defined the concept of an All Hazard Plan, and explained the benefits derived from such a plan.

Key Terms

Standard Operating Guidelines (SOGs)

Hazard Survey

Local Emergency Planning Committee (LEPC)

All Hazard Plan

Threat Analysis

Evacuation Plan

Mutual Aid Agreements

Mitigation

Preparedness

Response

Recovery

Continuity of Government

Continuity of Services

Response Phases

Increased Readiness

Pre-Impact

Impact Response

Sustained Emergency Phase

Review Questions

1. Explain the difference between a departmental Standard Operating Guideline and the EOP.
2. An All Hazard Plan will generally define response actions to types of incidents common to your response area based on identified hazards developed from the local area hazard assessment. What types of incidents should be listed?
3. Why should the EOP serve as a guide more than a directive to a response?
4. Define preparedness activities, and list some examples of this type of activity.
5. Explain the benefits of the EOC, and how the designation of an EOC will support ongoing incident operations.

Chapter   5

The Role of Planning During an Event

"A goal without a plan is just a wish."

Antoine de Saint Exupéry

Learning Objectives

After reading this chapter, you should be able to

- Identify the five key steps to be completed prior to developing your Incident Action Plan
- Define operational periods
- Understand how to effectively use the Planning P
- Identify Command and General Staff responsibilities during the planning process
- Understand operational priorities
- Develop a Planning Meeting agenda
- Understand when to create a written Incident Action Plan
- Create a SMART plan

You can find all forms referenced in this chapter in Chapter 6. The planning process during an emergency incident is no different than planning meetings used by large corporations when they are working on business projects. Generally, an American corporation creates a business plan to identify goals and objectives to be achieved. In the planning process for an emergency incident, you will need to do the same with the use of an Incident Action Plan. The beautiful thing about the planning process is that you can use it for everyday events such as a local parade or community festival and turn around and apply the same planning methodology to an emergency incident.

This chapter will guide you through the planning process and how to apply this to any situation. Understand that the key to an effective planning process is the use of the Incident Action Plan (IAP). This plan can be verbal or written, and this chapter will help you identify when to use a verbal plan and when you should consider a written plan.

Benefits of Good Planning

What are the benefits of good planning? Effective planning will provide both the management staff and operational personnel with guidance on what needs to be accomplished during your event. Notice I said *event*. The reason for this is that as you read through this chapter, you will see that the planning process can be applied to either an emergency or nonemergency event to achieve the same outcome: a successful event. Here are some of the things that effective planning will help you to achieve:

- Incident efficiency
- A common operating picture for all participating agencies
- A work plan
- Identification of what's done, where you are, and where you are going
- A road map to the successful fruition of an event

Five Key Steps to Take Before You Write Your Plan

When you are developing your IAP, whether verbal or written, you will want to take five key steps.

1. Identify what you want to do (ICS Form 202).
2. Identify who will be responsible for doing it (ICS Form 203).
3. Determine how to accomplish the identified tasks (ICS Form 204).
4. Define how units on the scene will communicate (ICS Form 205).
5. Determine what procedures on-scene resources follow if there is an injury (ICS Form 206).

Note: You can find these forms on this book's accompanying CD.

Unfortunately, developing your IAP is not as simple as answering these five questions and the completion of the five identified forms. It *may* be, but generally, as the event expands, so will the plan and the accompanying documentation. When developing your written IAP, you will need to identify your operational period. Generally, operational periods will run 4 to 24 hours depending on the type and nature of the event. The USFA and the Emergency Management Institute both use 12 and 24 hours as identified work periods to be used when creating a written IAP. The EQ Fire started as a verbal plan, followed soon thereafter with the first 12-hour written plan, which was 10 pages long. As the Apex event unfolded, the plan increased, as did the documentation. Copies of the IAP forms used for the Apex event appear on the accompanying CD.

The Planning P

The National Fire Academy, in conjunction with NIMS, has developed what they refer to as the **Planning P.** Figure 5-1 shows an example of the Planning P and how it applies to events and incidents.

By following the steps identified in the Planning P, you will be able to ensure that you cover all the points of the planning process in the appropriate order. The concept of the Planning P was initially created by the United States Coast Guard for use in oil spills and other large environmental emergencies that they were responsible for. The Coast Guard realized that they needed a guide to help them through their planning procedures. The Planning P was developed and is used as a guide to ensure that all areas of planning are covered for each operational period. You will note that the "P" continues in its form just as your planning process should. As you continue around the figure of the "P," you will continue to reassess your event, your objectives, and your available resources.

Operational Periods

During the planning process, you will need to identify your **operational periods.** An operational period is the identified work time for resources and identified objectives. It is the timeline for meeting the Incident Commander's expectations and objectives. The Incident Commander sets the length of time for an individual operational period based on objectives, available resources, and other considerations, which include the following:

- Safety of responders, victims, and other operating agency personnel
- Available resources on-scene versus en route resources that will have to be brought in
- The timeline needed to achieve certain tactical assignments
- Additional jurisdictions involved
- Weather and environmental conditions

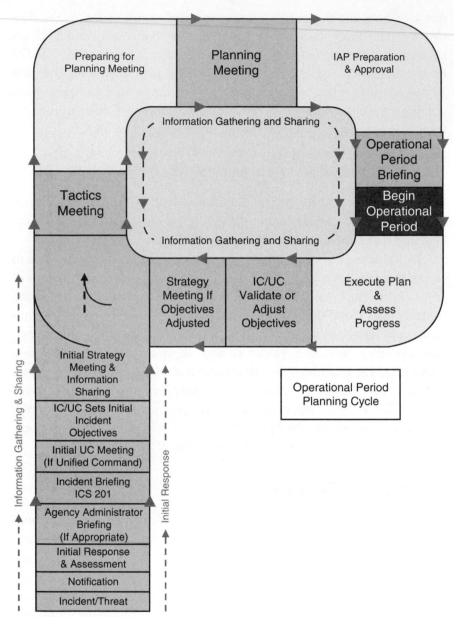

Figure 5-1 *Planning P applicability. Courtesy of the United States Fire Administration.*

Operational periods are generally set up around a 4-, 8-, 12-, or 24-hour window. This is generally determined during the Incident Command Strategy meeting.

Command and General Staff Responsibilities in Planning

Each player within the Command and General Staff will have certain identified responsibilities in the planning process. The positions and their duties follow.

Incident Commander (IC)
- Determines overall incident objectives and strategies for completing each objective
- Identifies procedures used for ordering resources
- Identifies procedures used for resource activation and mobilization
- Approves the completed IAP document

Safety Officer
- Reviews the identified objectives based on strategy and tactics, then determines safety considerations related to the same objectives
- Develops a safety statement to be part of the IAP that relates to the tactics used for accomplishing objectives

Operations Section Chief
- Assist the IC in determining strategies
- Develops tactics to accomplish objectives of the event
- Identifies work assignments and necessary resources

Planning Section Chief
- Develops status reports on the incident and the resources
- Develops a contingency plan—alternative plans to utilize in the event that primary plans fail to meet the determined objectives
- Oversees the planning process
- Develops the Incident Action Plan

Logistics Section Chief
- Identifies the resource ordering procedures
- Ensures that there is a system for the transportation of resources and other operational needs

- Validates that Logistics' resources can support the Incident Action Plan
- Places orders for any identified resources or other incident support materials based on event needs

Finance/Administration Section Chief

- Provides cost analysis of the incident objectives
- Ensures that the IAP is written within financial limitations as identified by the IC
- Evaluates the facilities, transportation needs, and other services that the event requires

Refer to Figure 5-2 for the flow of responsibilities. Before each operational period, all incident objectives should be assessed and necessary changes made as they are identified. The changes would address identified tactical changes, enhanced resource needs, and personnel requirements as determined by the incident objectives. Basically, the objectives will drive the tactics to be implemented, the determined amount of resources necessary to accomplish the tactics, and the number of personnel required to carry out the tactics. After this, you will need to hold a **Tactics meeting.**

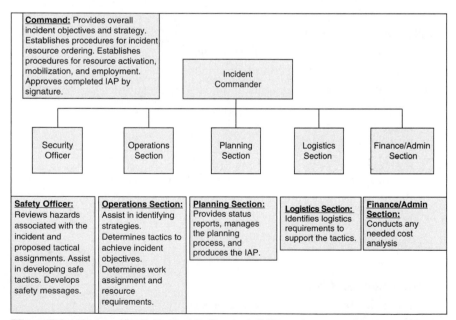

Figure 5-2 *Command and General Staff responsibilities. Courtesy of the United States Fire Administration.*

During this meeting, you will identify strategies and what necessary tactics to employ to accomplish the objectives of the event. Review what resources you have available and then evaluate whether you have the appropriate amount and type of resources to address the objectives. The Operations Section Chief should monitor all work being employed and ensure that the work is meeting the objectives identified.

Operational Priorities

Operational priorities should be documented during the Tactics meeting. Figure 5-3 identifies the three operational priorities to develop during the Tactics meeting: objectives, strategies, and tactics to use during the incident or event. Use the ICS 215 Form to document your results. This will then become part of the finished IAP.

Typing Resources

As you develop tactics, you will need to ensure that you have the appropriate resources on hand to implement and accomplish these tactics. To facilitate this task, NIMS has identified resources by type and capability. Resource managers use various resource inventory systems to determine availability of certain resources that may be available through public, private, or volunteer organizations. Generally, management systems

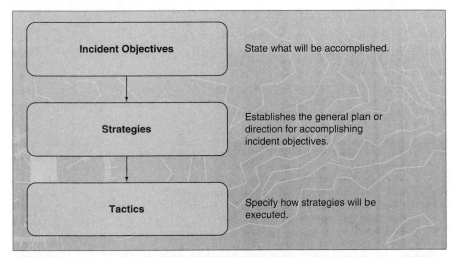

Figure 5-3 *Operational priorities: Objectives, strategies, and tactics. Courtesy of the United States Fire Administration.*

maintained at the local, state, and national levels track available resources. In order to make sure that specific resource needs are met, these agencies utilize the methodology of resource typing, which allows an agency to make requests for certain, specific needs and know that that resource need will be met. If you make your resource requests by type, it will save time, minimize errors, and give direction as to specifically what is needed. Resources should be described and identified by the following criteria:

- **Kind:** What the resource is; such as engine or water tender
- **Type:** Describes the size, capability, and staffing

You can find the complete typing list at: www.fema.gov/nims or on the accompanying CD.

Safety Analysis

After the Tactics meeting, you will need to perform a **Safety Analysis.** This analysis will help you identify the associated hazards that the event and the tactics implemented to secure the event present. The **Safety Officer** will develop this assessment and complete the ICS Form 215A to identify the safety concerns and what necessary actions or personal protective equipment will be used to mitigate those concerns.

Figure 5-4 shows proper documentation on a dummy form 215A. Once all the necessary information has been assembled, it is time for the **Planning meeting.** At this meeting, the Planning Section Chief will lead the discussion based on identified incident objectives and allow all Command and General Staff an opportunity to review and validate the plan as proposed by the Operations Section Chief. At the completion of the meeting, once all primary and alternative strategies have been discussed, the Planning Section Chief will identify which forms should be completed so that the final IAP document can be put together and duplicated for all personnel to use during the identified operational period. During the Planning meeting, the following documents should be displayed: incident objectives, a map of the area, the ICS 215, ICS 215A, and a meeting agenda.

Figure 5-5 shows an example of the planning meeting agenda that specifies who is responsible for each task. Remember, the individual who created one of the best plans ever implemented said, "Plans are nothing, planning is everything." That was General Dwight D. Eisenhower—the man responsible for planning D-day.

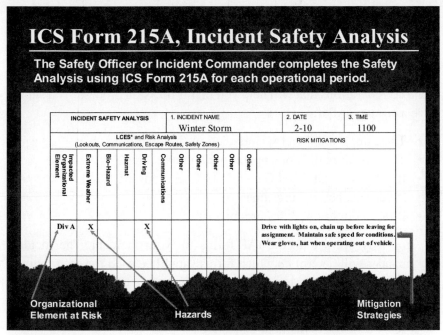

Figure 5-4 *ICS Form 215A—Incident Safety Analysis. Courtesy of the United States Fire Administration.*

Planning Meeting Agenda	Responsibility
Situation and resource briefing	Planning Section Chief
State incident objectives and policy issues	Incident Commander
State primary and alternative strategies detail tactical assignments, safety issues, and resource requirements	Operations Section Chief; Planning/Logistics Section Chiefs and Safety Officer contribute
Specify reporting locations and facilities	Operations Section Chief, Logistics Section Chief assists
Identify the resources, support, and overhead needed	Planning/Logistics Section Chiefs Logistics Section Chief places orders
Consider additional support requirements	Logistics Section Chief Planning Section Chief contributes
Discuss fiscal constraints, contracts, and claims	Finance/Administration Section Chief
Discuss safety issues not already covered, public information, and interagency liaison issues	Command Staff
Finalize, approve, and implement IAP	Planning Section Chief finalizes IAP; Incident Commander approves IAP; General Staff implements IAP

Figure 5-5 *Planning agenda. Courtesy of the United States Fire Administration.*

Deciding on a Written Plan

The question still remains: When should I have a written plan? Next are some considerations that will help you in this decision process. You should have a written IAP if:

- There are more than two jurisdictions involved.
- There is more than one operational period.
- If both Command and General Staff positions are filled.
- If your agency requires one.
- If the event involves hazardous materials.

You may not use all the IAP forms for your incident. The Incident Commander will determine which forms should be used for each individual incident. On smaller, less complex incidents, you can get by with using only the following forms:

- ICS Form 202—Incident Objectives
- ICS Form 203—Organization Assignment
- ICS Form 204—Assignment List
- ICS Form 215A—Incident Safety Analysis
- A map of the event area

These five forms will be the foundation of your initial IAP from which you can build upon as the event unfolds. The written IAP is a document that is compiled of a standard set of ICS forms and other supporting documents. It is the responsibility of the Incident Commander to identify just which forms will be used to make up an individual event's IAP. All the forms may not be used as they may not be applicable to the event or necessary in the conveyance of the Incident Commanders directions. However, generally the five forms listed previously will be the basic format for a written IAP. Using the forms, you will identify what needs to be accomplished, who will oversee the operations, who will complete the tasks, and how tasks will be accomplished safely. The map will identify where the operations will take place.

However, the sole use of these five forms is not a shortcut to less documentation if the event requires a more extensive IAP. The IAP needs to be applicable to your event in both size and functionality. Use what you need to make the document a success. The success of the event is directly related to the value of your IAP. Remember, this document is your game plan. You want it to address everything that you are trying to do and be a source that your team can use to guide them through the actions needed to bring your situation to a safe and successful end.

Using the SMART Method

Throughout this chapter we have referenced the development of objectives. Since this must be completed before developing tactics and carrying out operations, objectives should be clear and attainable in a defined time line. To ensure that you can meet your objectives, you may want to consider using the **SMART method.** Courtesy of the USFA, this method is used in developing objectives. It is simply an acronym used to provide guidance to the Incident Commander or to both Command and General Staff during larger events. Remember that all objectives are to be established based on three priorities: **life safety, incident stabilization,** and **property preservation.**

When you as the Incident Commander or a member of the staff have to develop objectives, you may wish to use this simple acronym as a guide to ensure that your objectives are in fact attainable. The definition of the SMART acronym is as follows:

S—Specific	Make sure that what you develop is unambiguous and specific to what you want to achieve.
M—Measurable	Make sure that accomplishments can be easily identified and can be measured.
A—Action	Write the objectives to identify what actions should be accomplished.
R—Realistic	Make sure what you want done is realistic. You can't move a mountain in one operational period, but you can get started.
T—Time	You should be aware of timelines. Identify what timelines you are operating within to make sure the time allotted is adequate for the required task.

Chapter Summary

In Chapter 5, we reviewed the need for and the benefits of planning at an emergency incident. We introduced the Planning P and how it can help you to cover all bases in the planning process. This chapter discusses the role of the Planning Section Chief as well as how the positions assigned to the general staff play into the planning process. It also discussed operational priorities and how to type resources along with Tactic meetings and Planning meetings.

The chapter specified when you should use a written plan and how to develop SMART objectives as part of the planning process. It also discussed the importance of well-defined objectives.

Key Terms

The Planning P	Planning Meeting
Operational Period	SMART Method
Tactics Meeting	Life Safety
Safety Analysis	Incident Stabilization
Safety Officer	Property Preservation

Review Questions

1. Define operational period. How long might it last?
2. What is the benefit of the Incident Safety Analysis?
3. Explain what the role of the Incident Commander is during the Planning meeting, and what his or her responsibilities are.
4. When should the Incident Commander consider the use of a written Incident Action Plan?
5. Explain the SMART method of identifying objectives.

Chapter 6

Incident Action Plans: How Do I Use Them?

"Create a definite plan for carrying out your desire and begin at once, whether you are ready or not, to put this plan into action."

Napoleon Hill

Learning Objectives

After reading this chapter, you should be able to

- Define the types of forms used to complete an Incident Action Plan
- Identify the forms that should be used at every event
- Identify what each form is used for and who is to complete the form
- Recognize the additional forms that can be used to manage larger events

The idea of an IAP is to serve as a road map for your emergency event. As we discussed in Chapter 5, the IAP may be verbal or written depending on the size of the event. Some individuals are hesitant to adopt written IAPs for managing their event, but it is not as complicated as it may first appear when looking at the documents that are generally used to complete the IAP.

Introduction to the Forms

The National Oceanic Atmospheric Administration (www.NOAA.gov), and NIMSonline (www.nimsonline.com) are only a few sites that provide downloadable IAP forms for field use. Many of these forms can be downloaded either as a Word document or Portable Document Format (PDF). The confusion starts when you look at the number of forms in each of these databases. Not to worry—all the available forms do not have to be used for every event. If you refer back to Chapter 5, you will see that the primary forms to complete for any event are as follows:

- ICS Form 202—What do you want to do? (Incident Objectives Sheet)
- ICS Form 203—Who will be doing the assigned tasks, and who is responsible for actions taken? (Organizational Assignment List)
- ICS 204—How will tasks be completed? (Assignment List)
- ICS 205—How will we talk to each other? (Radio Communications Plan)
- ICS 206—What do we do if there is an injury? (Medical Plan)
- ICS 215A—Generally required by OSHA personnel, this form will identify safety procedures to be implemented during the event. (Safety Analysis Form)

Apex Chemical Fire/Second 12-Hour IAP

We will use the IAP used for the second 12 hours of the EQ Fire for an example and explain what each page requires. At the end of this section, we will list the other IAP documents and how to complete them when needed.

IAP Cover Sheet

This is an optional document that is used to serve as a guide to the rest of your IAP, listing the forms used and providing space for additional optional documents. The four required items you must fill in for the cover sheet are as follows:

1. Incident Name—this could be a street location, city name, or the name of the event such as "Tropical Storm Diana" or "Fourth of July Celebration."
2. Desired timeline of the operational period (usually 4, 8, 12, or 24 hours).
3. Command Staff approval—on the NOAA documents, FOSC stands for Federal On-Scene Commander, SOSC stands for State On-Scene Commander, and RPIC stands for Responsible Party Incident Commander. However, you can substitute your own Command Staff titles in place of these if needed. For the Apex incident, we changed the NOAA document to list Incident Commander, Deputy Incident Commander, and Safety Officer. Each event will dictate what staff positions you use. The form is simply a tool; feel free to change it as needed to fit your event.
4. Individual who prepared the cover sheet.

See Figure 6-1 for an example of an IAP Cover Sheet filled out for Apex.

1. Incident Name	2. Operational Period to be covered by IAP (Date / Time)	IAP COVER SHEET
1005 INVESTMENT BLVD	From: 10/06/06-1000 To: 10/06/06-2200	

3. Approved by:

FOSC   CHIEF HARAWAY (FIRE) _2ʰᵈ_ _____

SOSC   CHIEF LEWIS (LEO) _____

RPIC   MICHAEL BOCHMAN (EMS) _____

_____   _____

_____   _____

INCIDENT ACTION PLAN

The items checked below are included in this Incident Action Plan:

☒   ICS 202-OS (Response Objectives)

☒   ICS 203-OS (Organization List) - OR - ICS 207-OS (Organization Chart)

☒   ICS 204-OSs (Assignment Lists)
One Copy each of any ICS 204-OS attachments:

     ☒ Map
     ☐ Weather forecast
     ☐ Tides
     ☐ Shoreline Cleanup Assessment Team Report for location
     ☐ Previous day's progress, problems for location

☒   ICS 205-OS (Communications List)

☒   ICS 206-OS (Medical Plan)

☐   _____

☐   _____

☐   _____

☐   _____

☐   _____

☐   _____

4. Prepared by:	Date / Time	
LT J WHITE	10/06/06	06.00

IAP COVER SHEET	June 2000

Figure 6-1 *IAP Cover Sheet. Courtesy of the Town of Apex.*

The next document you will find in the Apex Incident Action Plan is a map of the response area. This is very useful when outside agencies are coming to your aid. The map will help agency representatives identify where they are in regard to the event. You can obtain maps from local government planning departments, city engineer's office, county tax office, local forestry representative, online, and many other sources too numerous to list. If your incident involves a building or buildings, you may wish to include a floor plan during an incident as well. See Figure 6-2 for a map of Apex. The star denotes the location of the Town Hall.

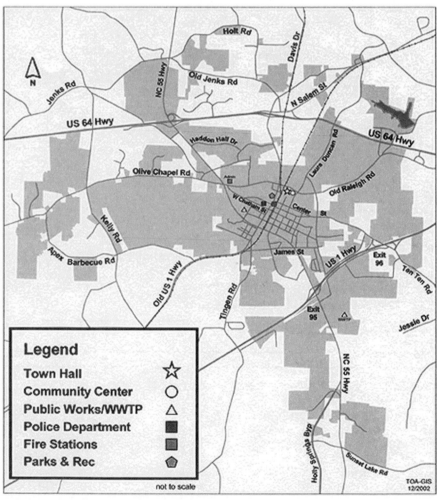

Figure 6-2   *Map of Apex response area. Courtesy of the Town of Apex.*

ICS 202—Incident Objectives

This form will list the overall and specific work period objectives as identified by Command and General Staff. You'll want to fill in the following fields in the ICS 202 form.

1. Incident name
2. Operational period
3. Overall objectives—these should be clear and concise, and will remain for the duration of the event.
4. Objectives specific to the operational period identified—these should be clear but short, well-defined objectives to be accomplished during the work period.
5. Safety Message—this field will identify known hazards related to the event and any safety concerns identified by the Safety Officer and/or Command. You will notice there is a space provided to identify where the site **Safety Plan** can be found; make sure there is such a plan and the location is listed. State and federal safety officials will want to see this if they are on the scene.
6. Weather—in this field you will either identify weather concerns or simply write in "see attached Weather Notice."
7. Tides and Currents—this field will apply when operating in coastal areas where changing tides can affect your operation.
8. Time of Sunset/Sunrise—in this field you would put in times relative to sunrise and sunset. This may be important if running night operations so as to indicate when lighting would be needed.
9. Attachments—simply check for and attach any other documents that would apply to your IAP.
10. Fill in the IAP preparer's name.

See Figure 6-3 for an example of an ICS 202 Form filled out for Apex.

ICS 203—Organization Assignment List

This form lists the organizational structure for the event and what units are assigned as well as what their assigned role may be. You do not have to use all positions. This form is usually prepared by the **Resource Unit Leader,** but could be done by another staff member if a Resource Unit Leader has not been assigned.

1. Incident name
2. Operational period

1. Incident Name	2. Operational Period (Date / Time)	INCIDENT OBJECTIVES ICS 202-OS
1005 INVESTMENT BLVD	From: 10/06/06-1000 To: 10/06/06-2200	

3. Overall Incident Objective(s)

Provide for the safety and welfare of all emergency responders and citizens

Provide care and shelters for any displaced citizens

Control and Eliminate any hazards associated with the incident

Conserve and protect any property associated with the incident

4. Objectives for specified Operational Period

Continue to provide for the safety of the responders and members of the general public
Continue to provide for the displaced citizens
Monitor the general area (Apex and surrounding) for possible expansion of the exclusion zone
Continue to restrict all air space of a 5 mile radius of 1005 Investment Blvd
Develop additional strategies dependent upon briefing from Haz Mat Branch Manager
Air Quality to continue to monitor air quality
DOT / LEO TO RESTRICT ALL ENTRY DUE TO STATE OF EMERGENCY

5. Safety Message for specified Operational Period

Command has had to be relocated for the 4th time. No chemical has been identified. There has been and still may be fire and has been explosions. ALL COMPANIES REMEMBER APPROPRIATE PPE PER YOUR JOB DESCRIPTION

SAFETY OF THE EMERGENCY RESPONDERS IS OF UTMOST IMPORTANCE

Approved Site Safety Plan Located at: COMMAND POST

6. Weather   See Attached Weather Sheet

7. Tides / Currents   See Attached Tide / Current Data

8. Time of Sunrise   **Time of Sunset**

9. Attachments (mark "X" if attached)

☐ Organization List (ICS 203-OS)	☐ Medical Plan (ICS 206-OS)	☐ Resource at Risk Summary (ICS 232-OS)
☐ Assignment List (ICS 204-OS)	☐ Incident Map(s)	☐
☐ Communications List (ICS 205-OS)	☐ Traffic Plan	☐

10. Prepared by: (Planning Section Chief)	Date / Time
Lt J White	10/6/2006 6:00

INCIDENT OBJECTIVES	June 2000	ICS 202-OS

Electronic version NOAA 1 0 June 1, 2000

Figure 6-3   *ICS 202 Form—Incident Objectives. Courtesy of the Town of Apex.*

3. Incident Commander and Staff
4. Agency Representatives—this field is used to list other assisting agencies and their representatives.

5 through 8. List your General Staff positions and other positions that may be filled as you organize the management structure of your event. These include Planning, Logistics, and Operation Sections. Simply place personnel names in the positions they fill.

9. List who prepared the document. You will notice that this space is on each form used.

See Figure 6-4 for an example of a completed ICS 203 Form.

ICS 207—Incident Organization Chart

This form is populated automatically when using the electronic format of NOAA but can be filled in by hand by simply transposing the information from your ICS 203. The ICS 207 simply places your assignments in a flow-chart format.

ICS 204—Assignment List

This form is completed for Divisions and Groups working at your event. It works well in identifying assignments when there are multiple groups or divisions being implemented. You will notice that the Apex IAP used an ICS 204 for the Recon Group, Fire Branch, Medical Branch, and LEO (Law Enforcement Officer) Branch. The following is the list of what to complete on this form.

1. Incident name
2. Operational period
3. Branch
4. List the name of the Group or Division assigned
5. Fill in the Operations Section Chief, Branch Director, and Division/Group Supervisor
6. Resources assigned this work period will fill in this blank. Under this field, you would identify the Team Leader, method of contact, number of persons assigned, and remarks such as special tasks or requirements. There is a box to be checked if you are using a separate 204A form.
7. The Assignment field provides a space for a descriptive detail of tactical assignments.
8. In the Special Instructions field, provide special instructions, such as safety messages or hazard concerns.

1. Incident Name	2. Operational Period (Date / Time)	ORGANIZATION ASSIGNMENT LIST
1005 Investment Blvd	From: 10/06/06-1000 To: 10/06/06-2200	ICS 203-OS

3. Incident Commander and Staff

	Primary	Deputy
Federal:	Chief Haraway (Fire)	
State:	Chief Lewis (LEO)	
RP(s):	Randy Moore	
Safety Officer:	WC1 Daryl Alford	
Information Officer	Michael Wilson	
Liaison Officer	Karl Hugench	

4. Agency Representatives

Agency	Name
Haz Mat	RFD Greg Bridges
Cary Fire	Captain Pope
Fairview	Gerald Atkins
FED EPA	CHRIS RUSSELL
Wake EM	Brian McFeters

5. PLANNING SECTION

Chief	LT J WHITE
Deputy	
Resources Unit	
Situation Unit	
Environmental Unit	
Documentation Unit	
Demobilization Unit	
Technical Specialists	

6. LOGISTICS SECTION

Chief	Ed Brinson
Deputy	
a. Support Branch Director	
Supply Unit	
Facilities Unit	
Transportation Unit	
Vessel Support Unit	
Ground Support Unit	
b. Service Branch Director	
Communications Unit	
Medical Unit	
Food Unit	

7. OPERATION SECTION

Chief	Alan Capps (Apex FD)
Deputy	

a. Branch I - Division/Groups

Branch Director	Haz Mat Greg Bridges
Deputy	
Division / Group	RRT 3 Fayetteville
Division / Group	RRT 4 Raleigh
Division / Group	
Division / Group	
Division / Group	

b. Branch II - Division/Groups

Branch Director	EMS 204 - Randy Moore
Deputy	
Division / Group	EMS 20 Decon Medical Monitor
Division / Group	471 transport at command
Division / Group	SRV1
Division / Group	
Division / Group	

c. Branch III - Division/Groups

Branch Director	
Deputy	
Division / Group	
Division / Group	
Division / Group	
Division / Group	
Division / Group	

d. Air Operations Branch

Air Operations Br. Dir	CECIL PARKER
Air Tactical Supervisor	
Air Support Supervisor	
Helicopter Coordinator	
Fixed Wing Coordinator	

8. FINANCE / ADMINISTRATION SECTION

Chief	BRUCE RADFORD
Deputy	
Time Unit	
Procurement Unit	
Compensation/Claims Unit	Robin Dawson
Cost Unit	

9. Prepared By: (Resources Unit)	Date / Time
	10/06/2006 01:17

ORGANIZATION ASSIGNMENT LIST	June 2000	ICS 203-OS

Electronic version NOAA 1 0 June 1, 2000

Figure 6-4 *ICS 203 Form—Organization Assignment List. Courtesy of the Town of Apex.*

9. Communications field—fill in radio, phone, and other contact numbers for assigned units.
10. Also be sure to fill in the person who completed the form.

See Figure 6-5 A through D for ICS 204 Forms (each assignment will have an ICS 204 form completed). As you review the ICS 204 forms provided in the figure, you will see that during the Apex event, each Branch assignment had an ICS 204 form completed. This is done so that each form would identify the specific referenced Branch and what its assignment was.

ICS 205—Radio Communications Plan

This document will list the event communications plan and includes the following:

1. Incident name.
2. Operational period.
3. List the type of radio system, channel(s) used, function of each assigned radio frequency, and remarks as needed.
4. Be sure to fill in who prepared the document.

See Figure 6-6 for an example of an ICS Form 205 filled out for the Apex incident.

ICS 206—Medical Plan

This form is used to provide information on the event's medical aid provisions, aid station locations, and transport services. The Medical Unit Leader completes this form, and it should also be reviewed by the Safety Officer. The fields that should be filled in on this form are:

1. Incident name.
2. Operational period.
3. List the name, location, and contact information for identified aid stations.
4. List the means of transportation used for medical transport.
5. List the names and locations of hospitals to be used for the event.
6. List any special medical emergency procedures pertinent to on-scene personnel.
7 and 8. List the name of the preparer and date of document with the Safety Officer's signature.

See Figure 6-7 for an example of an ICS 206 Form filled out for the Apex incident.

<table>
<tr><td>1. Incident Name</td><td colspan="2">2. Operational Period (Date / Time)
From: 10/05/06-2200 To: 10/06/06-1000</td><td>ASSIGNMENT LIST
ICS 204-OS</td></tr>
</table>

1. Incident Name	2. Operational Period (Date / Time) From: 10/05/06-2200 To: 10/06/06-1000	ASSIGNMENT LIST ICS 204-OS
3. Branch HAZ MAT	4. Division/Group RECON	

5. Operations Personnel

	Name	Affiliation	Contact # (s)
Operations Section Chief	CHIEF CAPPS	APEX FIRE	150*26*47309
Branch Director:	GREG BRIDGES	RALEIGH FIRE / HAZ MAT	
Division/Group Supervisor:			

6. Resources Assigned This Period "X" indicates 204a attachment with special instructions

Strike Team / Task Force / Resource Identifier	Leader	Contact Info. #	# of Persons	Notes / Remarks	
Raleigh Haz Mat / RRT 4	Capt Bridges		11		☐
Fayetteville / RRT 3					☐
					☐
					☐
					☐
					☐
					☐
					☐

7. Assignments

CONTINUE TO PROVIDE FOR SAFETY OF RESPONDERS
COMPLETE RECON OF INCIDENT SCENE
DEVELOP STRATEGY TO CONTROL AND ELIMINATE HAZARD DEPENDENT UPON FINDINGS OF RECON ASSIGNMENT
IMPLEMENT CONTROL AND ELIMINATION MEASURES

CONTINUE AIR QUALITY MEASURING
INITIATE WATER QUALITY MEASURING

8. Special Instructions for Division / Group

REMEMBER DEALING WITH AN UNKNOWN CHEMICAL(S) USE EXTREME CAUTION
USE APPROPRIATE PPE DEPENDENT UPON JOB DESCRIPTION

PAY ATTENTION TO CREW ROTATIONS DUE TO FATIGUE

WITH LEVEL "A" ENTRY REMEMBER BACK UP TEAM, DECON & MEDICAL MONITORING

9. Communications (radio and / or phone contact numbers needed for this assignment)

Name / Function	Radio Freq / System / Channel	Phone	Pager
CHIEF CAPPS / FIRE OPS	FIRE OPS 1		
CAPT BRIDGES / HAZ MAT	HAZ MAT OPS 1		

Emergency Communications

Medical FIRE OPS 1	Evacuation	Other

10. Prepared By (Resources Unit Leader)	Date / Time 10/6/06 6.00	11. Approved By (Planning Section Chief)	Date / Time 10/6/06 6:00
ASSIGNMENT LIST		June 2000	ICS 204-OS

Figure 6-5A *ICS 204 Form—Assignment List. Hazardous Materials Branch responsible for actual chemical containment. Courtesy of the Town of Apex.*

1. Incident Name	2. Operational Period (Date / Time)		ASSIGNMENT LIST
	From 10/06/06-1000 To: 10/06/06-2200		ICS 204-OS

3. Branch	4. Division/Group
HAZ MAT	RECON

5. Operations Personnel

	Name	Affiliation	Contact # (s)
Operations Section Chief:	CHIEF CAPPS	APEX FIRE	FIRE OPS 1
Branch Director:	CAPTAIN DAWSON	APEX FIRE	FIRE OPS 1
Division/Group Supervisor:			

6. Resources Assigned This Period   "X" indicates 204a attachment with special instructions

Strike Team / Task Force / Resource Identifier	Leader	Contact Info. #	# of Persons	Notes / Remarks	
APEX E2, E3, E4, E5, T1, L13	CHIEF CAPPS	R3,R1,T1,C1,B1,C5	29		☐
BAYLEAF 125	PRICE		2		☐
BOISE SPRINGS E842	CHRIS ATKINS		3		☐
PARKWOOD E611	RUDICILL		3		☐
CARY T7 L5	HINTON		8		☐
DURHAM CITY E 12	CURIA		5		☐
ANGIER BLACK RIVER					☐

7. Assignments

CONTINUE TO PROVIDE FOR SAFETY OF RESPONDERS
ASSIST MEDICAL GROUP AS NEEDED W/ PATIENT CARE
ASSIST HAZ MAT GROUP AS NEEDED
RESPOND TO ANY ODOR INVESTIGATIONS, PLUMES ETC IN THE AREA

8. Special Instructions for Division / Group

REMEMBER DEALING WITH AN UNKNOWN CHEMICAL(S) USE EXTREME CAUTION
USE APPROPRIATE PPE DEPENDENT UPON JOB DESCRIPTION

PAY ATTENTION TO CREW ROTATIONS DUE TO FATIGUE

9. Communications (radio and / or phone contact numbers needed for this assignment)

Name / Function	Radio: Freq. / System / Channel	Phone	Pager
CHIEF CAPPS / FIRE OPS	FIRE OPS 1		
CAPT DAWSON / FIRE BRANCH	HAZ MAT OPS 1FIRE OPS 1		

Emergency Communications

Medical	FIRE OPS 1	Evacuation	Other

10. Prepared By (Resources Unit Leader)	Date / Time 10/6/06 6.00	11. Approved By (Planning Section Chief)	Date / Time 10/6/06 6.00
ASSIGNMENT LIST		June 2000	ICS 204-OS

Electronic version NOAA 1 0 June 1, 2000

Figure 6-5B *Hazardous Materials Branch responsible for Fire Suppression. Courtesy of the Town of Apex.*

1. Incident Name	2. Operational Period (Date / Time)		ASSIGNMENT LIST
	From: 10/06/06-1000 To: 10/06/06-2200		ICS 204-OS

3. Branch	4. Division/Group
MEDICAL	

5. Operations Personnel	Name	Affiliation	Contact # (s)
Operations Section Chief:	CHIEF CAPPS	APEX FIRE	FIRE OPS 1
Branch Director:	RANDY MOORE	WAKE EMS	EMS OPS 1
Division/Group Supervisor:			

6. Resources Assigned This Period   "X" indicates 204a attachment with special instructions

Strike Team / Task Force / Resource Identifier	Leader	Contact Info. #	# of Persons	Notes / Remarks	
EMS 20	ALAN FOSTER		2		☐
471	EDDIE WOODALL		2		☐
SRV1	MATT HINSLEY		1		☐
					☐
					☐
					☐
					☐

7. Assignments

CONTINUE TO PROVIDE FOR SAFETY OF RESPONDERS
RESPOND TO ANY MEDICAL RESPONSES IN AND ROUND THE EFFECTED AREA
***** ANY RESPONSES ARE TO BE A "RAPID EXTRICATION" PER EMS 204

8. Special Instructions for Division / Group

REMEMBER DEALING WITH AN UNKNOWN CHEMICAL(S) USE EXTREME CAUTION
USE APPROPRIATE PPE DEPENDENT UPON JOB DESCRIPTION

PAY ATTENTION TO CREW ROTATIONS DUE TO FATIGUE

9. Communications (radio and / or phone contact numbers needed for this assignment)

Name / Function	Radio: Freq. / System / Channel	Phone	Pager
CHIEF CAPPS / FIRE OPS	FIRE OPS 1		
RANDY MOORE / MEDICAL	EMS OPS 1		

Emergency Communications

Medical	EMS OPS 1	Evacuation		Other	

10. Prepared By (Resources Unit Leader)	Date / Time 10/6/06 6 00	11. Approved By (Planning Section Chief)	Date / Time 10/6/06 6 00
ASSIGNMENT LIST		June 2000	ICS 204-OS

Electronic version NOAA 1 0 June 1 2000

Figure 6-5C *Medical Branch responsible for providing medical care to first responders and also coverage of any medical calls in the affected area. Courtesy of the Town of Apex.*

1. Incident Name 1005 Investment Blvd	2. Operational Period (Date / Time) From: 10/06/06-1000 To: 10/06/06-2200	ASSIGNMENT LIST ICS 204-OS
3. Branch LEO	4. Division/Group	

5. Operations Personnel

	Name	Affiliation	Contact # (s)
Operations Section Chief:	CHIEF LEWIS	APEX PD	
Branch Director:	CAPT STEPHENS	APEX PD	
Division/Group Supervisor:			

6. Resources Assigned This Period   "X" indicates 204a attachment with special instructions

Strike Team / Task Force / Resource Identifier	Leader	Contact Info. #	# of Persons	Notes / Remarks	
					☐
					☐
					☐
					☐
					☐
					☐
					☐

7. Assignments

SECURITY FOR EFFECTED AREA
PERIMETER AREAS ARE 64, 55, 1010
RESPOND TO REQUEST FOR SERVICES TO ASSIST WITH EVACUATIONS
SECURE ANY CRIME SCENES ASSOCIATED
MAINTAIN TRAFFIC CONTROL THROUGH OUT AREA

8. Special Instructions for Division / Group

ALL LEO PERSONNEL IN THE EFFECTED AREA OBTAIN AND USE PPE
APPROPRIATE PPE WILL BE A RESPIRATOR AND TYVEK SUIT
ANY MUTUAL AID LEO WILL PROVIDE LIKE OR SAME PPE PRIOR TO DISPATCH

9. Communications (radio and / or phone contact numbers needed for this assignment)

Name / Function	Radio: Freq. / System / Channel	Phone	Pager
LEO	PD2		

Emergency Communications

Medical	EMS OPS 1	Evacuation	Other

10. Prepared By (Resources Unit Leader)	Date / Time 10/6/06 8:49	11. Approved By (Planning Section Chief)	Date / Time 10/6/06 8:49
ASSIGNMENT LIST		June 2000	ICS 204-OS

Electronic version NOAA 1 0 June 1, 2000

Figure 6-5D *Law Enforcement Branch responsible for perimeter security, crime scene preservation, traffic control, and maintenance of evacuated area. Courtesy of the Town of Apex.*

1. Incident Name 1005 INVESTMENT BLVD		2. Operational Period (Date / Time) From 10/05/06-2200 To 10/06/06-1000			INCIDENT RADIO COMMUNICATIONS PLAN ICS 205-OS

3. BASIC RADIO CHANNEL USE

SYSTEM / CACHE	CHANNEL	FUNCTION	FREQUENCY	ASSIGNMENT	REMARKS
WAKE COUNTY 800 NOTE CHANGE	MUTUAL AID 3	FIRE BRANCH ACTIVITIES		MAIN INCIDENT CHANEL	
WAKE COUNTY 800 NOTE CHANGE	MUTUAL AID 2	STAGING & LOGISTICS		STAGING & LOGISTICS	
WAKE COUNTY 800	EMS OPS 1	MEDICAL OPERATIONS		MEDICAL OPERATONS	
WAKE COUNTY 800	EMS OPS 2	MEDICAL STAGING		MEDICAL STAGING	
WAKE COUNTY 800	PD 2	LEO OPERATIONS		LEO OPERATIONS	
		HAZ MAT RECON		HAZ MAT RECON	
CARY 800	MUTUAL AID 2	CARY FIRE		CARY FIRE	
WAKE COUNTY 800	FIRE OPS 4	APEX EOC		APEX EOC	

4. Prepared by: (Communications Unit)			Date / Time

INCIDENT RADIO COMMUNICATIONS PLAN	June 2000	ICS 205-OS

Electronic version NCAA 1.0 June 1, 2000

Figure 6-6 *ICS 205 Form—Incident Radio Communications Plan. Courtesy of the Town of Apex.*

1. Incident Name	2. Operational Period (Date / Time)	MEDICAL PLAN
1005 INVESTMENT BLVD	From: 10/06/06-1000 To: 10/06/06-2200	ICS 206-OS

3. Medical Aid Stations

Name	Location	Contact #	Paramedics On site (Y/N)
OLIVE CHAPEL	OLIVE CHAPEL SCHOOL	RED CROSS	
TURNER CREEK	TURNER CREEK ELEMENTATY	RED CROSS	
COMMAND POST	COMMAND POST	471	
SRV 1	TO ASSESS ANY MEDICAL RESPONSE IN AREA	SRV1	

4. Transportation

Ambulance Service	Address	Contact #	Paramedics On board (Y/N)
STAGING	AS NEEDED AFTER 471		

5. Hospitals

Hospital Name	Address	Contact #	Travel Time		Burn Ctr?	Heli-Pad?
			Air	Ground		
WAKE MED CARY						
DUKE HEALTH RAL						
REX						

6. Special Medical Emergency Procedures

CONTINUE TO MONITOR FROM ILL EFFECTS FROM EARLY EXPOSURE
WITH ANY NAUSA, VOMITING OR HEAD ACHE CONTACT AID STATION

7. Prepared by: (Medical Unit Leader) Date / Time	8. Reviewed by: (Safety Officer) Date / Time

MEDICAL PLAN	June 2000	ICS 206-OS

Electronic version NOAA 1.0 June 1, 2000

Figure 6-7 *ICS 206 Form—Medical Plan. Courtesy of the Town of Apex.*

Weather Assessment

Figure 6-8 shows a weather assessment provided by the National Weather Service for the day of the Apex fire. Having this type of information in your IAP will provide up-to-date weather information for on-scene personnel.

ICS 215A—Incident Action Plan Safety Analysis

The last form used in the Apex IAP is the ICS 215A—Incident Action Plan Safety Analysis. This document is used to list specific safety concerns and required safety guidelines for all personnel operating at your event. It is simple to complete—just identify your safety mitigation actions and then enter the Division or Group that it applies to. You will note that the provided document is formatted for a wildland scenario, but you can simply write in your event-specific safety concerns and identify safety mitigations accordingly. For a filled-in version of the ICS 215A, see Figure 6-9.

As I mentioned at the beginning of the chapter, there are other IAP forms available, and each has a defined application depending on the type and nature of your incident. On the CD that accompanies this book, there is a copy of all the IAP forms in a Word format. These documents can also be found and downloaded electronically from www.nimsonline.com. The following are the remaining documents with a short synopsis of their individual use and application.

- ICS 207—Organizational flow chart to be filled in identifying all staff positions.
- ICS 209—Incident Status Summary—indicates status of all resources on scene at the event.
- ICS 211—Incident Check-In List—used by agencies responding to your event. This document is generally completed prior to arriving on scene or can be completed during check-in at staging. The form identifies what resources and personnel an agency is arriving with and their capabilities.
- ICS 213—General Message Form—this document is used to send messages between staff positions and Command staff, such as resource requests and other event needs. The document can also be used to send event messages from Command to the EOC or other informational centers that are a part of your event.
- ICS 214—Unit Log—This document is used by Team Leaders to track their assigned personnel and activities during a work period.

Hazardous weather conditions for

Apex, NC

Enter Your "City, ST" or zip code [   ] <> [Go]

2 products found:

Hazardous Weather Outlook

```
HAZARDOUS WEATHER OUTLOOK...CORRECTION
NATIONAL WEATHER SERVICE RALEIGH NC
450 AM EDT FRI OCT 6 2006

NCZ007>011-021>028-038>043-073>078-083>086-088-089-070845-
PERSON-GRANVILLE-VANCE-WARREN-HALIFAX-FORSYTH-GUILFORD-ALAMANCE-
ORANGE-DURHAM-FRANKLIN-NASH-EDGECOMBE-DAVIDSON-RANDOLPH-CHATHAM-
WAKE-JOHNSTON-WILSON-STANLY-MONTGOMERY-MOORE-LEE-HARNETT-WAYNE-
ANSON-RICHMOND-SCOTLAND-HOKE-CUMBERLAND-SAMPSON-
450 AM EDT FRI OCT 6 2006

THIS HAZARDOUS WEATHER OUTLOOK IS FOR PORTIONS OF CENTRAL NORTH
CAROLINA.

.DAY ONE...TODAY AND TONIGHT
SCATTERED THUNDERSTORMS WILL OCCUR TODAY AND EARLY THIS EVENING. A
STORM OR TWO MAY PRODUCE PENNY SIZE HAIL AND WIND GUSTS AROUND 45
MPH.  RAINFALL AMOUNTS THROUGH MIDNIGHT TODAY WILL VARY FROM LESS
THAN A HALF INCH OVER THE SOUTHERN COUNTIES TO AROUND AN INCH OVER
THE FAR NORTHERN PIEDMONT AND NORTHERN COASTAL PLAIN COUNTIES WITH
A FEW SPOTS RECEIVING BETWEEN ONE AND TWO INCHES. A NORTH WIND WILL
INCREASE LATER TODAY AND HOLD STEADY BETWEEN 10 AND 15 MPH WITH
GUSTS AROUND 25 MPH. FINALLY...MUCH COOLER TEMPERATURES ARE EXPECTED
TODAY WITH HIGHS IN THE 60S NORTH TO THE 70S FAR SOUTH.


.DAYS TWO THROUGH SEVEN...SATURDAY THROUGH THURSDAY
COOLER WEATHER CONTINUES SATURDAY WITH AFTERNOON TEMPERATURES
REMAINING BELOW 60 DEGREES OVER THE NORTH. RAIN WILL GRADUALLY END
FROM NORTH TO SOUTH SATURDAY AFTERNOON AND SATURDAY NIGHT. THE NEXT
THREAT FOR THUNDERSTORMS WILL BE WEDNESDAY INTO THURSDAY. A FEW OF
THE STORMS MAY BE STRONG...MAINLY ON THURSDAY.

.SPOTTER INFORMATION STATEMENT...
SPOTTER ACTIVATION IS NOT ANTICIPATED FOR TODAY NOR TONIGHT.
HOWEVER SPOTTERS ARE ENCOURAGED TO REPORT ANY THREATENING WEATHER
SUCH AS HAIL...WIND GUSTS STRONG ENOUGH TO DOWN TREE LIMBS OR
RAINFALL AMOUNTS IN EXCESS OF AN INCH TO THE NATIONAL WEATHER
SERVICE.

$$

WSS
```

Short Term Forecast

```
SHORT TERM FORECAST


NATIONAL WEATHER SERVICE RALEIGH NC
835 AM EDT FRI OCT 6 2006

NCZ024-025-041-061400-
ORANGE-DURHAM-WAKE-
INCLUDING THE CITIES OF...CHAPEL HILL...DURHAM...RALEIGH
835 AM EDT FRI OCT 6 2006

.NOW...
A BAND OF MODERATE TO OCCASIONALLY HEAVY RAIN SHOWERS WITH
ISOLATED THUNDERSTORMS WILL CONTINUE TO MOVE EAST ACROSS THE
TRIANGLE THROUGH 10 AM. RAINFALL AMOUNTS FROM ONE QUARTER TO ONE
HALF INCH WILL BE POSSIBLE UNTIL THE RAIN TAPERS OFF FROM WEST TO
EAST SHORTLY BEFORE NOON. NORTH NORTHWEST WINDS AT 5 TO 10 MPH
WINDS WILL INCREASE TO 10 TO 15 MPH BY NOON. TEMPERATURES WILL
HOLD STEADY IN THE MID 60S.

$$

MWS
```

Figure 6-8 *Hazardous Weather Outlook. Source: National Oceanic and Atmospheric Administration.*

INCIDENT ACTION PLAN SAFETY ANALYSIS

LCES* Analysis of Tactical Applications
Lookouts Communications Escape routes Safety zones

1. Incident Name: 1005 INVESTMENT BLVD
2. Date: 11/6/06
3. Time: 16:00 - 22:00

LCES Mitigations:
- UNKNOWN CHEMICAL
- PROPER PPE & BACK UP
- MEDICAL MONITORING
- DECON ESTABLISHED PRIOR TO ENTRY
- UNKNOWN CHEMICAL
- PROPER PPE
- ALL RESPONSES RAPIDEXTRICATION
- UNKNOWN CHEMICAL/PROPER PPE
- UNKNOWN CHEMICAL
- PPE AVAILABLE AS NEEDED

Column headers (rotated): Division/Group, Overhead Training, Individual Training, Mid Rope Training, Anchor Points, Previous Casualties (Location, Wind driven), Rehab Potential, Hazard Material, Transportation, 1 hr +, Communications, Structure Protection, Other Risk Analysis, Other Risk Mitigations

Division/Group entries: HAZ-MAT AT, EMS, LINE, LEO

Prepared by (Name and Position): John White, P. DANNIE

ICS 215A

Figure 6-9 ICS 215A Form—Incident Action Plan Safety Analysis. Courtesy of the Town of Apex.

- ICS 215—Operational Planning Worksheet—This form is used as part of the assignment planning for an operational period. The document will aid in identifying what each unit has and what they may need for the assigned work period.
- ICS 216—Radio Requirements Worksheet—This form is used to identify radio communications needs for each Division/Group assigned. The form breaks down each Division/Group's radio needs and identifies their assigned equipment for communications and frequencies allocated.
- ICS 218—Support Vehicle Inventory—This form identifies vehicle resources assigned to an event and includes their size, capability, fuel needs, etc.
- ICS 220—Air Operations Summary—This is the form to be used when an Air Operations Branch is assigned and air units are being utilized. The form summarizes your air support inventory on scene and any needs they may have.
- ICS 221—Demobilization Checkout—This form is used to check out resources as you start to demobilize. Each unit will complete one of these forms and once each Section Chief from your General Staff assignments has completed their part on the form, the unit can return to service and clean up from the event.
- ICS 224 and 225—Personnel Rating Forms—These documents are used as evaluation forms for personnel who worked at or were deployed to the event.
- IMT 1—Incident Management Team Checklist—This document can be used as a checklist of things the Incident Management Team needs to complete.
- IRSS Check-In Form—This document is used to check in arriving resources.

The aforementioned forms are simply additional documents that are available to you for use during an event. Understand that they are not necessarily required nor are they applicable to every event you respond to. They are provided on the accompanying CD for your review. As the Incident Commander or member of an Incident Management Team, you should know what documents are needed based on the type of event you are responding to.

Chapter Summary

Chapter 6 detailed how to use an Incident Action Plan and the documents that complete such a plan. This chapter discussed each of the main plan documents in detail with examples given to aid you in understanding how to use each form. In addition, it identified which forms should be used for all events and listed other forms that staff may use on larger events to aid in the management of the event. The chapter integrated how the IAP documents were used in the Apex event and how the documents served as a written event business plan. I think if you look at what was identified on each of the documents, you will see that the document creates written guidance for what each person is supposed to do and how they are to complete their tasks. The documents demonstrate how to fill in the ICS forms, which then give you a written IAP for your event.

Key Terms

Safety Plan   Resource Unit Leader

Review Questions

1. Identify which forms should be used for all events when creating the written IAP.
2. Define what the ICS 204 form is used for.
3. Explain why it is important to include a current weather statement in the IAP.
4. The ICS 215A identifies safety considerations for the event. Who is responsible for completing this document, and why is it important?
5. What is the importance of the ICS 216 form?

Chapter 7

Resource Management

Learning Objectives

After reading this chapter, you should be able to

- Identify the five principles of resource management
- Identify the considerations related to managing event resources
- Identify who is responsible for ordering resources during the event
- Differentiate between single-point and multi-point ordering
- Understand the check-in process

Resources at an incident are there for one purpose: to help you accomplish the incident objectives. To make sure that on-scene resources are utilized to successfully accomplish identified objectives, they must be organized and utilized properly. The effective management of on-scene resources will be one of the most important tasks during your incident. During the Apex incident, Command and General Staff had to manage a large number of resources to accomplish the many tasks at hand. Early identification of what types of resources would be needed was instrumental to the effective, desired outcome. With the Apex basically split in half, it was identified early in the incident that resources would need to be in place to both respond to the chemical fire incident and also other normal situations occurring in the unaffected parts of town. This was the driving factor for ordering large numbers of resources early on. Once resources were on the scene, they were assigned based on objectives identified by Command and tactical needs as identified by the Operations Section. Resources not immediately utilized were staged for future use or for replacement of existing resources. An example of this was the ordering of a second Hazardous Materials Team. During the firefighting/mitigation efforts, the Hazardous Materials

Branch Director realized that existing hazardous materials personnel working in the Hot Zone would have to be replaced on a regular basis due to
fatigue. With this in mind, the Hazardous Materials Branch Director requested
Logistics to order a second team early so that they would be in place to relieve
currently assigned personnel.

The Five Principles of Resource Management

During a large event, whether planned or not, you must ensure that you have
sufficient resources to respond effectively. According to NIMS, there are
five principles of resource management (see Figure 7-1). They are:

1. Advance Planning—It is imperative for organizations to work together
   to develop plans for managing and using resources.
2. Resource Identification and Ordering—Organizations should use standard procedures to identify, order, mobilize, dispatch, and track resources.
3. Resource Categorization—Resources need to be categorized by size,
   capacity, capability, skill, or other characteristics to make resource ordering and dispatch more efficient.

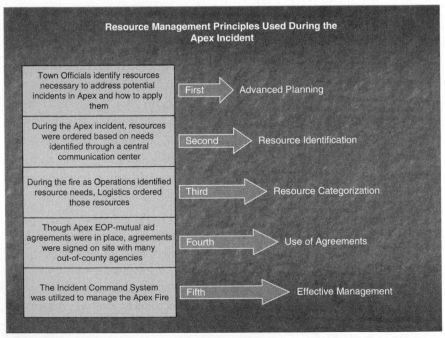

Figure 7-1 *Resource management principles applied.*

4. Use of Agreements—Mutual aid agreements should be established for resource sharing during large scale events.
5. Effective Management—Proven practices should be used to perform key resource management tasks.

Additional Resource Management Considerations

There are several considerations that must be taken into account when managing resources at an incident. These considerations are safety, personnel accountability, managerial control, adequate reserve resources, and cost of resources.

- **Safety**—Resource management should always include ensuring the safety of:
  1. Responders to your event/incident
  2. Anyone injured or threatened by the event/incident
  3. Any volunteer groups involved
  4. Any news media or other member of the general public who is on the scene
- **Personnel Accountability**—Your resources should be accounted for at all times. All supervisory personnel in the ICS structure should know where their assigned resources are at all times—this is known as **unity of command organizational structure.**
- **Managerial Control**—Performance and adequacy of the current IAP must be assessed and adjusted continually. Resource managers at all levels must continuously assess the effectiveness of the objectives in the IAP and adjust them as needed.
- **Adequate Reserve Resources**—You should always be anticipating incident needs and be ready to adjust resources as the incident dictates. Always make sure there are reserve resources in staging so as not to overtax existing units. This was evident at the EQ Fire in Apex when the original Command Staff was still working after 26 hours with no relief. It should be understood that as with any event, emergency responders learned many lessons. In the case of the Apex incident, adequate resources were in place to relieve both operational personnel and Command Staff, so this example from the Apex incident does not suggest that resources were not in place, they just were not utilized initially. Command, just like Operational Staff, should be

replaced in regular intervals to ensure that all personnel are fresh and capable of carrying out their assigned tasks.

- **Cost of Resources**—Since cost is always a major consideration in any large incident, resources should be ordered and used in the most cost-effective manner possible. Ensure that the Finance Section records cost, prepares cost summaries and cost estimates, and recommends cost-saving measures whenever possible.

Getting Started

The first task is to determine the kind, type, and quantity of resources needed to achieve the incident's objectives. One way to do this is to use the ICS 215 Form—Operational Planning Worksheet. This document (part of a written IAP) will identify the resources needed to implement the strategies and tactics of the event.

Ordering and Tracking Resources

What responsibilities are assigned with regard to the resource ordering process? Both Command and General Staff have specific assigned responsibilities with regard to resource ordering activities. Ultimately, the Incident Commander has the final approval for the ordering of resources used at an incident; however, during large incidents the Incident Commander may not have time to review every single order. Routine orders and the approval of those orders can be delegated to Logistics. If the specific resource request is not routine, has an increased cost associated with it, or requires activating outside agency responses, then the Incident Commander may want to review that order.

Ordering activities and the associated responsibilities are assigned as listed in the following bulleted list:

- Command Section—Develops objectives for the event/incident and approves all resource orders and the demobilization of those resources.
- Operations Section—Responsible for developing tactical assignments and oversees the resources needed to accomplish incident objectives as outlined by Command.
- Planning Section—Responsible for tracking resources allocated and any shortages of needed resources.
- Logistics Section—Responsible for ordering resources as needed.
- Finance Section—Responsible for paying for resources and tracking resource costs.

All of the previously listed activities are identified as part of the ordering and tracking process based on the fact that all sections of the Command structure play a major part in either identifying resources necessary for an incident, actually ordering those identified resources, tracking what has been ordered, approving orders, and paying for the cost associated with those orders.

Command typically approves all resource orders; however, the Incident Commander, Logistics Section Chief, or Supply Unit Leader can be authorized to place orders as well. Resource ordering can be accomplished by one of the following methods depending on the size of the event:

- **Small Events**—The Incident Commander may approve resource needs at the scene and relay these needs to the designated ordering point for that responsible jurisdiction.
- **Larger Events**—An Incident Commander may decide to set up **single-point ordering.** Under this process, the responsibility of ordering resources is done through a single designated dispatch center or EOC. If a Logistics Section has been established, then all requests are directed to Logistics who will then make the official resource order to the single-point agency designated. Single-point ordering is generally the preferred method to reduce duplicate orders, point-of-contact confusion and overordering. However, some instances, such as needing resources from multiple sources, needing resources from agencies other than the home agency, or needing special resources such as private companies, requires ordering from more than one place. An example of this would be what occurred during the Apex incident. Initially, all resources were being ordered through Wake County 911 via the Wake County Emergency Management Office. As the incident expanded, requiring more resources than were available through Wake County, the Logistics Section had to make contact with other outside agencies. This was accomplished by communicating resource needs through the North Carolina State EOC. Additional law enforcement resources utilized during the Apex incident were requested through their respective dispatch centers. Under these circumstances, you may have to implement **multi-point ordering,** where resource orders are placed through multiple or different points and even the private sector. This should only be done if absolutely necessary since this type of ordering places additional workloads on the incident personnel. It also requires more coordination to avoid lost or duplicated orders.

When ordering resources through single-point or multi-point ordering, you should make sure you communicate the following information to the respective agency from which resources are being ordered:

- Incident name
- Order/request number
- Date and time of order
- Quantity, kind, and type of resource being requested
- Special support needs
- Where responding resources should report to
- Whether resources are needed now or later (at a planned time)
- Communications needs
- Your name and title
- Phone number or radio call channel

As part of your resource ordering procedure, you will need to have a tracking process that identifies the resources' status. As ordered resources arrive on scene, they will need to check in. This necessitates some form of perimeter control. This will enable you to maintain better resource account-ability, control access to the scene, and ensure responder safety by providing a safe and secure work area.

Resource Check-In

Let's explore the check-in process. There are several areas within the command structure to monitor resources. They are as follows:

- The Incident Commander.
- The Planning Section Chief.
- The Resource Unit—this unit can be activated under the Planning Section Chief to organize and monitor check-in of arriving resources.

Resources can check in at any one of the following places as determined by one of the aforementioned management units: Incident Base, Base Camp, and **Staging Area,** at the ICP through the Resource Unit, or at the **helibase.** The helibase is the only area where air resources, such as a helicopter, could land and check in. In the field, the Staging Area would be the primary place for check-in to occur, but large events such as a wildland fire where one may have numerous camp operations established, resources may check in at that assigned camp location. Wherever check-in is performed, the ICS 211 Form should be used to document check-in of arriving resources.

Examples of Resources Ordering and the Check-In Process

In Figure 7-2, you will see the Apex Logistics Chief meeting with Wake County Emergency Management to determine resources needed for the EQ fire. The resource ordering process had to be implemented early into the Apex incident. Initially, the fire in Apex escalated from a single alarm response to a general alarm, skipping the second and third alarm. At the same time, Command Staff requested additional mutual aid units from neighboring departments, the City of Raleigh Hazardous Materials Team, EMS, and additional law enforcement units. This initial request was the equivalent of 8 engine companies, 3 ladder companies, 5 EMS units, 16 police units, and 3 hazardous materials team units. A Staging Area was established as resources moved to the fourth Command Post. Almost immediately, the Logistics Section Chief set up Level 2 Staging, and Command established an alternate Staging Area at the alternate EOC location. Command realized as the chemical plume expanded over a large area of the town, affecting more citizens and eventually cutting Apex in half, that a separate contingent of resources would have to be ordered and maintained at the alternate EOC to respond to emergency calls in the unaffected areas of town.

Figure 7-2 *Here, the Logistics Section Chief is coordinating arriving resources at the assigned check-in point and Staging Area, which was adjacent to the Command Post. Courtesy of Lee Wilson, photographer, Raleigh Fire Department.*

As resources continued to arrive, it became evident that an accountability process needed to be implemented as part of the initial check-in process. Initially, a firefighter was assigned to record units and personnel as part of the accountability process as soon as individual resources checked in. However, with resource needs escalating, the Resource Unit was established as part of the Planning Section to coordinate check-in with the Staging Area Manager. As units checked in, the Staging Area Manager would relay that information to the Resource Unit where the resource type, number of personnel, and personnel identification would be recorded. Figure 7-3 shows a firefighter who was assigned to maintain accountability of all on-scene resources for the EQ Fire.

At the height of the EQ Fire, resources such as apparatus, police units, and EMS units were staged in the parking lot of a vacant shopping center. This same shopping center is where the Command Post was relocated for the *fifth* time. Within the shopping center was a vacant grocery store, which was used to house responders. In the grocery store, a rehab area was set up in addition to hygiene facilities and a feeding area. By starting the resource tracking process early in the incident, each time the Command Post was required to relocate, it was easy to ensure that all on-scene resources were accounted for. To understand how important this is, simply look at the

Figure 7-3 *Accountability was initially tracked as units checked into the Staging Area during the Apex incident. Courtesy Lee Wilson, photographer, Raleigh Fire Department.*

Figure 7-4 *The Staging Area located adjacent to the fifth Command Post location. All on-scene resources were parked in a manner that would keep them in a readiness state to respond as soon as they were requested. Courtesy of Lee Wilson, photographer, Raleigh Fire Department.*

number of personnel that had to be moved when the wind direction changed, forcing the fifth movement of the Command Post, Staging Area, and Media Area. That relocation required accounting for 300 firefighters, over 200 police and EMS personnel, and 10 outside agencies that were on the scene. Apex officials also moved 54 news and media agencies during this relocation.

Figure 7-4 shows the Staging Area where equipment for response to the EQ Fire was held awaiting an assignment. Notice how the fire apparatus are lined up in a readiness state to respond as necessary.

Chapter Summary

In Chapter 7, you read about the five principles of resource management. The chapter also discussed what to address when ordering resources for an event. It also listed individual ordering responsibilities of both the Command and General Staff positions and evaluated single-point and multi-point ordering as well as when to use each type of ordering. It is important to track resources and maintain accountability of arriving resources and assigned personnel. The chapter gave some examples of how resource management was used during the Apex incident, including how to establish staging areas and maintain personnel accountability. Resource management is so important for you to understand since having the right resources and enough supporting resources for responders will determine much of an event's outcome. If you do not have the proper resources, you cannot expect personnel to accomplish their assigned tasks. At the same time, personnel cannot function without food, water, and proper hygiene support. The use of resource management will aid in an efficient, effective support operation.

Key Terms

Unity of Command   Multi-Point Ordering
   Organizational Structure   Staging Area
Single-Point Ordering   Helibase

Review Questions

1. When ordering resources for an event, your organization should use standard procedures to identify, order, mobilize, dispatch, and track resources. Define those procedures as identified in this chapter.
2. Define the difference between single-point and multi-point ordering and when you should use each one.
3. Explain where resources can check in when arriving at an incident.
4. What are the responsibilities of the Planning Section Chief in the resource ordering process?
5. Explain the resource tracking process and why it is important when ordering resources.

Chapter 8

The EQ Fire in Retrospect Through the Eyes of the Incident Commander

"Always have a plan B, plan A may not work."

Mark Haraway

Learning Objective

After reading this chapter, you should be able to

- Identify all strategies and concepts that are taking place as the event unfolds

Up to this point in the book, you have read about large-scale emergency situations and how NIMS was designed to guide agencies through these situations. This book defined the ICS and how this system can be used as a management tool to effectively organize responding agencies. You should have a good understanding of how the ICS is designed to expand as your incident grows into a complex situation requiring more resources. This book also discussed resource management principles as well as how to plan for such situations. You have been introduced to an incident complex and defined Area Command. To these alternative organizational models, you have been given some options as to how you may opt to organize complex incident management. We will now take you back to October 5, 2006, to the large hazardous materials release and subsequent fire that expanded into a very large incident and literally shut down a town for several days. You will get to see how the management principles you have just read about were used effectively to manage the many resources that had to be brought in to respond to this situation.

The chapter contains a vivid first-person depiction of the EQ Fire as seen through the eyes of the Incident Commander. I will take you back to the night of October 5, 2006. You will be inside the driver's seat for this event, facing the problems of that night and witnessing the decision making necessary to bring this event to a close. You will feel the pressure from the political leaders of the community, the fears of the citizens, and the questions from not only local but national media.

First 911 Call

As with any emergency, first responders have to be notified. In the following section, the initial 911 communications have been transcribed so that you can see how this incident started to unfold. Remember, your 911 telecommunicator is key to the Incident Command System since they are generally the first to receive the call.

First 911 Call at 2136

911 Operator: What is the location of your emergency?
Caller: Uh, over in Apex off of Schefflin Boulevard and Investment, there must be a chlorine leak over there, uh, a real, real strong chlorine smell and you can see the haze throughout the area, where it's uh, in the area where you can smell it at.
Call ended at 2137.

Second 911 Call at 2138

911 Operator: What is the location of your emergency?
Caller: Yeah, I was on Industrial Boulevard, down by Dream Sports Center in Apex, and I think one of the buildings over there has a fire or a chemical spill, there's a huge, uh gas cloud over the road there.
911 Operator: Yeah, they are aware of it and they are getting someone on the way.
One alarm was dispatched at 2138 October 5, 2006.
911 Operator: Odor investigation on Schefflin Rd and Investment Boulevard.
Apex Fire Department: Apex Engine 1 en route—2138, Apex Engine 3 en route—2138.
Apex Fire Chief: Apex, Car 1 will be en route to Investment Boulevard— 2139.
911 Operator: Apex to units en route to Investment Boulevard. The second caller from Dream Sports Arena has called reporting a strong smell of chlorine in the area.
Apex Fire Department: Engine 3 copies Apex.

Figure 8-1 *Site plan of the Apex industrial corridor. Courtesy of the Town of Apex GIS Department.*

What you have just read is the initial dispatch and radio traffic from Apex 911 and the first responding fire units. To give you an understanding of the area that units were responding to, see Figure 8-1, an aerial map of the location where callers were reporting the "cloud." You will see that the area has several commercial buildings and borders a residential neighborhood.

The first engine arrived on the scene at 9:42 p.m., but told dispatch they were stopping five blocks out because they couldn't get in the industrial park—they had been met by a large vapor cloud. I was listening to the initial radio traffic from the first arriving units stating that they could not get into the industrial park due to a large vapor cloud, and I realized this was more than a simple odor investigation. I generally don't get in a hurry too often, but based on what I was hearing, I had to that night. When I arrived shortly after the first engine on scene, I was amazed at what I saw.

You could not see the street lights, the building lights, the parking lot lights, or anything else in the industrial park. All I could see was a white cloud towering 30 feet, moving across the corridor, and already seven blocks into a residential neighborhood. The odor of chlorine was prevalent and pungent. Of course, we immediately went into a general alarm response, which in Apex activates the units not already in service as well as mutual aid units from two neighboring departments. We called all off-duty fire personnel back in. I had a radio in one hand and a cell phone in the other. I called dispatch and gave her a page-long list of people I needed as soon as possible. At the same time, on my cell phone, I was calling (and waking up) Brian McFeaters of the Emergency Management Office in Wake County. I said "Brian look, are you listening to the radio? No? I need you in Apex. We have a leak. I don't know where it's coming from, but I need you to activate the Reverse 911 system."

The first thing I thought of was that the Code Red System that I was so proud of hadn't been turned on yet. It was an automated system that the Apex Town Council had just purchased that allowed us to notify all our residents by telephone, cell phone, mobile phone, or text messaging. You see, earlier in the day I had had a conversation with Town Manager Bruce Radford, and we signed and prepared the paperwork to mail in to activate the Code Red System. The Code Red System is a notification system that was more inclusive than the Reverse 911 system. It could notify all numbers listed regardless of phone systems to include cell phones and play a recorded message that I can activate from my office or cell phone regardless of my location. Ironically, the last thing we discussed was that it was a great idea to implement the system in case anything ever happened. Little did I know how the evening would unfold.

However, I knew that Wake County had a Reverse 911 system that we have access to, so Brian started activating that. I told him two things needed to happen. First, we needed people under the cloud to shelter in place. Second, the rest of the downtown district, 22 blocks, needed to move to the community center. Luckily, we have a very large community center in town that is our designated primary shelter facility.

We sent one recon team in. They couldn't get through the cloud. When we sent a second team in, they had to go around the cloud and walk into the hazard area from the back side of the facility. To do this, they had to go through the parking lot of White Oil Company, which is a next door neighbor to EQ. They got in right about the same time I was talking to Captain Bridgers from the Raleigh Haz-Mat Team explaining about the vapor cloud and the potential surroundings and concerns we had. The Town Manager was also there now, as well as the Police Chief. We had already started the door-to-door evacuations with the six police officers we had on duty, while they called in the rest of the police force.

The second recon team had a brand new lieutenant on it; he had been there for only six months. I had already grabbed Captain Dawson, who was on duty, and gave him two radios and my cell phone and said to monitor all of them, and if anything important happened to tell me about it. I basically assigned Captain Dawson as my aid to answer radio and telephone traffic. I told Captain Dawson, "You stay close." As I was telling him this, I could hear "Engine 1 to Command," on the radio. Captain Dawson answered it. The Lieutenant said, "I need to talk to the chief." Dawson responded, "Stand by." The Lieutenant came back and said, "We don't have time to stand by. It's EQ." Well, my heart sunk and my ears tuned in, because I knew that was a problem.

EQ was the worst possible business in that area that could be on fire. Before I could get the microphone in my hand, the lieutenant came back and said, "No, be advised, it's on fire! No change that—it's through the roof!"

Well, he had no more than finished his call and the first explosion went off (refer back to Figure 1-2). It shook the ground and produced a giant fireball and mushroom cloud. The cloud was 150 feet high and resembled an atomic bomb explosion. Captain Bridgers exclaimed, "We're too close!" That was an understatement. We moved immediately to the Command Post at the top of a hill, which was about five blocks away. We stayed there for 10 minutes. In that 10-minute window, the manager of EQ arrived. I asked him where his manifest was. He told me that the inventory manifest was in his office. I then asked him if he knew what he had on site at the facility. He said, "Sure. We have large quantities of flammable liquids, combustibles, oxidizers, contaminated mercury, contaminated lead, poisons, and pesticides."

As the explosions continued, we decided to relocate to the second Command Post. We then tried to move a third time to the YMCA, but we couldn't stop there. We were seeing explosions in the rearview mirror, so we just drove right by. Finally, we stopped a quarter mile out at the Department of Corrections facility and set up our fourth Command Post. We had already upgraded the response to include public safety agencies from most of western Wake County, which included the Cary Fire Department, Cary EMS, Fairview Fire Department, Holly Springs Fire Department, Western Wake Fire Department, City of Raleigh Fire Department (including their

Haz-Mat Team), and law enforcement units from Cary, Raleigh, Wake County Sheriff's Department, and the North Carolina Highway Patrol. Of the 23 EMS units that are on duty 24/7 in Wake County, we had 16 of them in staging. We weren't really sure what we were going to be faced with, and we knew that we would not only have to deal with the fire and its effects but also the unaffected portions of Apex that would still require emergency services. I was trying to run Command from the back of my vehicle (see Figure 8-2). However, by this time, we had set up Unified Command consisting of the Apex Police Chief Jack Lewis, EMS Chief Nicky Winstead, and myself. I also had Wake County Emergency Management as part of the Command Staff. My vehicle was just too small for an event of this magnitude. I decided to request that Raleigh Fire Department send their Mobile Command Unit.

Law enforcement was already assisting with the evacuation at this point. Based on our information from Haz-Mat, we extended the evacuation zone to a mile in all directions. The Apex City Hall is located next to the community center, which is our town's shelter. We currently had people going there for shelter. Fire Station No. 3 is our EOC. Immediately when we went to the general fire alarm, I called in my off-duty Battalion Chiefs. Battalion Chief Lawrence Carter called me, and I told him to first get the EOC set up, and second, get another shelter opened up.

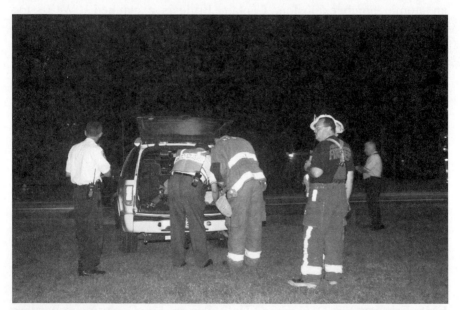

Figure 8-2 *Apex Car 1 being utilized as a Command Post. Courtesy of Lee Wilson, photographer, Raleigh Fire Department.*

While I was at the Command Post trying to get things organized, brief Command Staff, and set up the media area, Battalion Chief Carter called me on my cell phone. He said, "Hey chief, the plume is all over the shelter area. I've got 450 people—what are we going to do with them?" I told him to move the evacuees to Olive Chapel Elementary School, which was west of town. I told Command Staff what was going on. Chief Carter called a second time and said, "Hey chief, we just put personnel in full gear and air packs so we could get the equipment out of Station 1. The station is full of the chemical cloud that is already downtown. And it's encroaching on the police station." The police station houses our 911 call center. From that location, there was only four more blocks to the EOC. So I pulled the Command Staff together, talked to the Town Manager, and found out that the Mayor was at the shelter (see Figure 8-3). I told the Battalion Chief to take everyone to a new shelter. Chief Jack Lewis sent police officers to assist with getting the evacuees to Olive Chapel Elementary. Olive Chapel Elementary is the school farthest to the west of the event, just outside of the Exclusionary Zone. I said to my staff, "Look, you're going to move the EOC to the school. Get as much equipment quickly out of the station and get the guys out of harm's way and let's go."

Figure 8-3   *The briefing of the Town Manager and Police Chief on the situation and the fact that we have started to move evacuees to a new shelter location. Courtesy of Lee Wilson, photographer, Raleigh Fire Department.*

   We had 300 firefighters, 16 EMS units, 5 Haz-Mat resources, and state resources including the Department of Environmental and Natural Resources, Air Quality, and Water Quality on the scene. The school board was also involved because we had to address the issue of the school being used as a shelter and how to proceed with school the following day. Raleigh's command vehicle arrived on the scene, and all staff were moved to this unit to set up operations. Little did I know that before this event would end, the Command Post would be the only government infrastructure left in town.

   Once Command Staff was moved into the Raleigh command vehicle, the first planning meeting was held to prepare for event briefings (see Figure 8-4). We had already set up a media area at this point. The media area was adjacent to the command facility in a neighboring building's parking lot. For this particular event, the media was our best friend. We had set up a Medical Branch and had two paramedics assigned to contact citizens sheltered in place in the Exclusionary Zone. These medics were doing wellness checks on these people to ensure that they were okay. We had police officers route alerting or public addressing through their vehicles' intercom system to advise citizens of the evacuation and where they should go. As soon as we set up the media area and gave them something they could broadcast, they interrupted all local programming and started disseminating

Figure 8-4   *Raleigh's command vehicle arrives. Courtesy of Lee Wilson, photographer, Raleigh Fire Department.*

information on this fire. Our Public Information Officer (PIO) was briefing the media every hour and kept the updates rolling. For each update, they were interrupting the normal television programming. They really centered their focus on the event and keeping the public informed, telling people how to shelter in place, where to go if they were able, what routes to use to get out of the Exclusionary Zone and where emergency shelters were located.

We had branched out our organizational structure, which went from the top of the page to the bottom of the page. Our organizational management structure had been determined early on in the event. As the situation expanded, so did our management structure.

My first briefing with the City Manager and the Mayor was at the back of the mobile Command Post. I told them, "Look, based on that glow you're seeing, I have considered a city block gone to the fire and that's being optimistic." I went on to explain that we were evacuating everything east of Highway 55 and discussed what town facilities had already been lost. The Mayor's chin dropped. The Manager said, "What do you mean?" I said, "You have to understand—this is EQ." Realizing the magnitude of the situation, Command advised the Mayor to declare a state of emergency for Apex. At that point, the only thing we knew about the facility from the plant manager was that there were pesticides, poisons, contaminated mercury, contaminated lead, oxidizers, and a bulk amount of volatile fuels on the premises. And they couldn't give us any breakdown of what the actual chemicals were. When we got the **manifest,** it was three and a half pages long. The chemicals were identified only by an alphanumeric code, which had to be researched to determine what the codes stood for. When we finally decoded what chemicals the facility contained, it was clear how delicate the situation really was (see Figure 8-5).

The biggest potential problem we were facing was the dangerous materials that were stored in surrounding facilities. White Oil, a 400,000-gallon storage facility for petroleum, is right next door to EQ. Three hundred feet behind it is Apex Cabinet Shop, a commercial manufacturing facility with lacquers and all kinds of other flammables on the site. Across the street is Apex Metal Works, a welding shop, and just up the street is Apex Gymnasium. There was plenty of fuel around to keep the fire going for a while. Amazingly, even with the fireballs and mushroom clouds from chemical containers that were going off, the fire never did spread to the other facilities. None of those 55-gallon containers landed in one of the other buildings. They landed in the adjacent parking lot and in the street,

```
1-4 Naphthalenedione Reactive Materials
Nicotene and Salts
Solid Waste—Flammable, Ignitable
Corrosive
Copper Cyanide
Mercury
Potassium Cyanide
Spent Cyanide/Plating Bath Solutions
Methyl Ethyl Ketone Peroxide
Halogenated Solids
Lead
Non-Halogenated Solids
Tetrachloroethylene
Acetic Acid Ethyl Esther
Petrachlorophenol
Halogenated Solvents—Degreasing
Spent NLH Solvent
Chromium
Selenium
Benzene
Cadmium
Ethylene Dibromide
Tri, tetro-pentachlorophenol
Phenol
Hydrozine-RT
Thioperoxiydicarbonicdioxide-H2N
Epinephrine
Arsenic
Carbon Tetrachlorine
Methyl-chloroform
Endrin
Chloroform
Barium
Vinyl Chloride
Waste Waters
2,4-D
Arsenic-Oxide
Trichloroethylene
Benzene 1.1
Chlordane
Chlordane Alpha and Gamma Isomers
Silver
Methane-Dichloro
Formaldehyde
Carbaryl
Creosote
Methyl-Ethyl Ketone
Sodium-Cyanide
Waste Water Treatment Sludges
```

Figure 8-5   *EQ chemical list.*

but they just didn't hit anything that would cause the fire to spread past the warehouse.

Every time we had a briefing, we had to pull the three Incident Commanders into the Unified Command Center. We also had to meet with the elected officials. We didn't bring them into the mobile Command Post. All the decisions were being made in there by me, the police chief, and the Unified Command Staff. We'd just go out and report to them every so often.

We had to evacuate Fire Stations No. 1 and No. 3, we shut down town hall, and we moved the EOC several times. One thing we were able to do once we identified how fast this thing was rolling was to evacuate Public Works, not of people but equipment. I asked that the Town Manager contact our Public Works Director and have him get his staff to move all of their equipment to the alternate EOC. He took every dump truck, every piece of heavy equipment, and moved it off to EOC just so we would have it. By daybreak, our plan was to go in and hopefully recon the site, find out how much fire was left, do some kind of containment, and try to pick up the pieces.

As the winds grew, it became obvious we were going to lose our 911 center. Apex has its own dispatch center where we dispatch EMS, fire, and police without having to go through the countywide center. We were very fortunate that we were able to partner with the neighboring group, North Carolina Canine Emergency Response Team (NCCERT). NCCERT functions as a partner community agency to the Apex Fire Department, and they have a mobile communications vehicle. It has all the different types of radios, satellite phones, ham radio, VHF, UHF, and 800 MHz— everything one would need to communicate. They brought this truck out for us to use. Although it's actually housed at one of our stations, we have to get their people out to operate it, and they have licensed ham radio operators. NCCERT arrived with their communications unit and we moved our dispatchers in there and put together a patch between Wake County and Apex: We didn't miss a call. The call center operated out of the communications vehicle for three days. We tied and patched in to the Wake County center, routed the calls, and routed radio traffic. This event was initially assigned six radio channels, which allowed the Command Staff to organize each agency on a channel, thus making overall radio communications more efficient. As an additional safety precaution, we had to shut down the airspace over Apex for five square miles. This was particularly important because Apex sits in the middle of the approach zone of Raleigh-Durham International Airport. We shut down the air space, and that created a nightmare on the East Coast for flight plans. Then we shut down the railroad. The switchyard is in downtown Apex. We called to inform them of the situation, and they said, "You got what? . . . A big chemical fire." "Where's it at? . . . All over Apex." The freight train stopped dead in its tracks. It blocked every crossing. That was a godsend for us because we had everybody out and they couldn't get back in. In addition, CSX shut down the entire CSX line for three days. Amtrak also had to be rerouted.

Of course, the Red Cross was assisting with shelters by this time. In addition to occupying the Olive Chapel Elementary School, we opened up

Figure 8-6   *Evacuation shelter—This is a view of citizens who were awaken in the middle of the night and moved to several shelters until they arrived at this final shelter— number 3, Olive Chapel Elementary School. Courtesy of Mike Legeros, photographer.*

a second shelter and were looking to open a third (see Figure 8-6). We were using schools for all the shelters; there are nine schools in Apex. The chemical cloud had now spread to fill in most of our entire evacuation area. Apex is surrounded by highways 55, 64, and U.S. 1. Those highways basically made up our Exclusionary Zone. We had highway patrol, law enforcement from six neighboring jurisdictions, plus the Wake County Sheriff's Department, Durham County Sheriff's Department, and the DOT involved. Every point of entrance from one of those major highways into Apex was shut down. That became our **control point.**

As soon as the North Carolina Department of Environmental and Natural Resources got there and set up their mobile lab, they were able to go around the Exclusionary Zone and set up monitoring pods that tied into the mobile lab so they could start taking air samples during the event. By the end of the third day, they had taken 252,000 air samples around the Exclusionary Zone and the perimeter!

On top of all the obstacles we faced with the EQ Fire and the surrounding hazardous areas, the wind changed six times during the event. We had a member of the Emergency Management Staff on site at the Command Post, and they were receiving constant weather updates from the airport.

They came out one time during a briefing and said the winds had moved again. I said I thought we had a wide enough zone. But they said, "You don't understand where it's moving to. It is heading toward the nursing facility." In Apex we have a nursing facility with bedridden patients who cannot care for themselves. So now, not only are we trying to evacuate most of the town, we also had to try and figure out how we were going to move these individuals out. We had to reroute ambulances because the nursing facility was on the far side of town. We routed them through Cary, a neighboring city, just to get into Apex, probably going 30 minutes out of the way. We ended up using the 16 ambulances we had in staging, the 5 not in service, every wheelchair van we could find, plus 3 mass transit busses. Basically we set up a separate branch under EMS for this special evacuation. By this point, we decided Unified Command was a great concept, but we needed to take it one step further. So we carried it into an incident complex, which allowed us to take our General Staff positions and make more of them. Suddenly, I had a Fire Operations Section; a Haz-Mat Operations Section; an EMS Branch to take care of the other half of the city that wasn't affected by the incident; an EMS Branch for treatment, triage, and transport for nothing but this nursing home; and a Law Enforcement Branch. Basically, we moved 17,000 citizens in three hours; in four hours, we displaced and found shelter and housing for the nursing home patients so they could stay in bed and maintain their level of care.

It was about 2:30 a.m. on October 6 that we ran into another problem as we were moving citizens. Some of them inadvertently went through the plume, so we were getting calls from our three main hospitals now saying they were getting walking, self-deployed wounded, or people who were showing signs of contamination. So, we created another Area of Command outside of our main incident complex because I had to send Raleigh, Cary, and Morrisville Fire Departments to all three of these hospitals to set up technical decon outside of every emergency room (see Figure 8-7).

If there's one thing that I swore up and down I'd never do, it's to put my people at risk. I know it's inevitable to some degree with the job we do, but there's one thing that the Fire Chief really doesn't want to hear, and that's when your own people have become victims. Chief Lewis and I were in the Command Post when a law enforcement officer came in and said, "We've got a problem. We've got three officers showing signs of contamination." I was thinking maybe they were having a little trouble breathing, but I told them we'd go see what was going on. I told Chief Winstead of

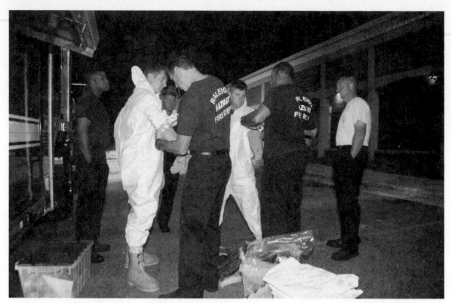

Figure 8-7   *Firefighters suiting up and preparing for one of the decontamination sites at a local area hospital. Courtesy of Lee Wilson, photographer, Raleigh Fire Department.*

Emergency Medical Services, "Look you've got the Command Post right now. We'll be right back."

The police officers were bleeding from their noses, had tears also coming out of their eyes, *and* they were having trouble breathing. Some were experiencing skin rashes. So we got decon set up and started running them through. We had three EMS units set up in staging just prior to this. We brought all three of them over. We figured three people, three units. Before we could get them through decon, four more police officers were showing with the same symptoms. Then there were three more. Then came a firefighter. In all, we sent 13 responders to the hospital. The worst of it was that we didn't know what was wrong with them. We couldn't tell the doctors what they'd been exposed to. We had no idea what was in this plume. My biggest concern was that I had a firefighter whose only exposure to the plume was during recon, when he was in full gear and with breathing apparatus. He had also been through decon after that brief time, and he was still showing severe signs of exposure. See Figures 8-8 through 8-10.

After the responders went to the hospital, I had to make another report to the manager: "Hey, we just sent 13 first responders to the hospital suffering from severe chemical exposure." The Town Manager then had to tell the Mayor the news that we now had some of our own personnel going to

Figure 8-8 *Initial decon being set up for one of the exposed police officers. Courtesy of Lee Wilson, photographer, Raleigh Fire Department.*

Figure 8-9 *Decon line is growing as more first responders fall victim to the chemical cloud. Courtesy of Lee Wilson, photographer, Raleigh Fire Department.*

Figure 8-10 *Decon worker awaits another responder to come through the line.*
Courtesy of Lee Wilson, photographer, Raleigh Fire Department.

the hospital with severe chemical exposure. Losing these first responders created yet another problem for us, because like I said, Apex is a small town. We have only 6 police officers on duty at any given time. Chief Lewis had called his next 6, so that was the 12 we sent to the hospital, plus a firefighter. The patrol cars that these 12 officers were operating were out of service because they were all contaminated, so Haz-Mat locked them down and said "Look we don't know what the exposure is, but if the officers are contaminated so are the vehicles. These cars are staying right here." The rest of the patrol cars, which other officers don't take home, were also down in the contaminated zone. So all of a sudden we lost a large portion of the police officers as well as police cars. Chief Lewis requested additional law enforcement units. We needed everything. We couldn't get to any of our equipment or anything else on the law enforcement side, and we had to maintain this Exclusionary Zone. We had law enforcement agencies from six neighboring jurisdictions, highway patrol, city police departments, and the Sheriff's Office respond and assist.

At 4:30 a.m. on October 6, our weatherman at the airport contacted Command and said, "You aren't going to believe it." By this point I didn't believe much of what was happening since around every corner was another seemingly insurmountable obstacle. Everybody was looking to me for answers, and I had about run out of answers. Our weatherman said a low pressure system had developed just west of Apex. I stood there a minute. After all, this was well after midnight, and I was tired and worn out. In my experience, low pressure systems usually meant a hurricane. And he said "Well, yeah. It's pretty much a hurricane; it just isn't where it ought to be." And I said, "Well, yeah, who in the world invited it? I've got enough problems. I don't need a hurricane to include in the mix." Sure enough, it moves right in, took a turn, and made the wind we were trying to work with turn 180 degrees. That took the cloud with all the contaminants back toward the Command Post, our staging area, and our EOC. So Logistics and Planning had to get together quickly and come up with a plan that wasn't chaotic and would allow us to move all the resources that we had pulled together. This time we commandeered a whole shopping center that was located 2.2 miles from the scene. And within a one-hour window, we moved our alternate EOC, all our staging, all our Command, to that point—it was pretty much a convoy. We started under the cover of darkness, moved up the street, and found a route that we could actually get through. You know you have a large-scale incident when the sun comes up, you step out of the command vehicle, and emergency vehicles fill every marked-off parking spot in the shopping center. The shopping center not only provided the optimal location for a Command Post and Staging Area for the large number of resources but also gave us plenty of empty buildings to set up as a rehabilitation area for our personnel. The large building allowed Logistics to set up cots for responders as well as restroom facilities and an area for food and hydration. See Figures 8-11 and 8-12.

We had all our state agencies, federal agencies, the EPA, the Atlanta Strike Team, and C-Tech, which is a third-party toxicology group that works with the EPA. Alcohol, Tobacco and Firearms was on the scene as well as the Centers for Disease Control and Prevention. I was finding out what every acronym in the state and federal government stood for because everyone showed up, and I was meeting them all. They were as nice as you can get, great to work with. Every one of them that came in was there, and said they were ready to work with us, and that they'd do anything they could to help us.

Finally, we got to the point where we could make entry into the Exclusionary Zone. During entry, we identified three working fires still in the facility. We came up with a plan where we weren't going to allow any reentry of civilians until we could completely put the fire out. EQ stepped forward; they brought us an industrial firefighting team from Arkansas. We worked with them, tied them in to our Haz-Mat Team and our

Figure 8-11   *In the Staging Area, firefighters wait for orders to be assigned. Uncertainty of the situation and knowing the exposure of fellow responders didn't deter the responders. Courtesy of Lee Wilson, photographer, Raleigh Fire Department.*

Figure 8-12 *Rehabilitation area where responders rested and awaited their assignments. Courtesy of Lee Wilson, photographer, Raleigh Fire Department.*

firefighting force. And over the next 14 hours worked to extinguish what was left of this facility fire.

The fire was finally out at 12:28 a.m. on Saturday morning, October 7. From that point, I handed off the lead Command role to Chief Lewis, who then came up with a reentry plan. We maintained Unified Command but with Chief Lewis in charge of the repopulation of Apex while I coordinated the investigative efforts with the SBI and ATF. We opted to do a basic plan of reentry because we had taken care of the whole town (see Figure 8-13). We started at the farthest point of the perimeter, bringing them back in one phase at a time, right up until right around the industrial park, which we kept closed for three more days until we could ensure that there was no contamination.

While Chief Lewis was busy with the reentry, I was working closely with the Wake County Fire Marshal, SBI, and ATF to determine how this fire started (see Figure 8-14). At the same time, cleanup contractors were arriving, and these efforts also had to be organized so as not to conflict with the ongoing investigation (see Figure 8-15). While on the site with investigators, I was finally able to see what had occurred and how our efforts were effective in controlling this incident. The Haz-Mat Team had built earthen berms prior to the firefighting effort, which controlled potentially harmful run-off (see Figure 8-16). All that remained was the burnt-out chemical containers and the remains of the facility (see Figure 8-17).

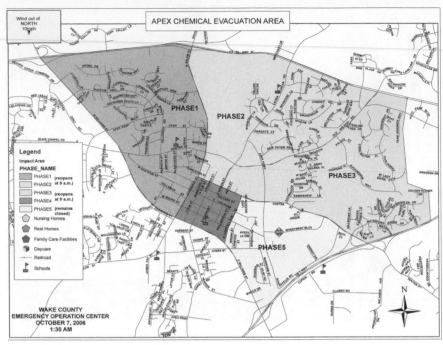

Figure 8-13 *5-Phase Re-Entry Plan for the Apex incident. We would repopulate based on geographic region with those in closest proximity to the fire being allowed to enter last. Courtesy of the Town of Apex Planning Department.*

Figure 8-14 *Aftermath—Investigators look over the site after the fires are out and the building is gone. Courtesy of the Apex Fire Department.*

Figure 8-15   *Cleanup. Courtesy of the Apex Fire Department.*

Figure 8-16   *Containment efforts—The sand berms were built around the site to contain water runoff. Courtesy of the Apex Fire Department.*

Figure 8-17   *Chemical drums as they were found. A lot of the tops were blown off, and some were bulging, ready to explode. Courtesy of the Wake County Fire Marshal's Office.*

Chapter Summary

Chapter 8 placed you in the driver's seat for what was one of the largest chemical fires in North Carolina in the past 25 years. You were faced with the situation, the decisions to be made, and the questions that had to be answered. This chapter hopefully gave you an opportunity to see what you could be faced with as an Incident Commander in charge of large events affecting your city. Was everything done correctly? It very seldom ever is. No one is perfect, but the key to success is not perfection but the capacity to manage the event, the personnel, and the resources needed to successfully bring it to an end.

Key Terms

Manifest   Control Point

Review Questions

1. What is the difference between EPCRA and RCRA as it related to the EQ event?
2. How many people were forced to evacuate during the EQ event, and how did the town provide shelter for the displaced citizens?
3. Why did the Lead Spokesperson for Command transition to the Police Chief at the end of the event?
4. How many times did the Command Post relocate and why?
5. What federal resources were called upon for assistance at the EQ Fire?

Chapter 9

The Aftermath: Reentry and Cleanup

> *"Learning is not attained by chance; it must be sought for with ardor and attended to with diligence."*
>
> *Abigail Adams*

Learning Objectives

After reading this chapter you should be able to

- Identify how Apex coordinated the reentry of citizens to the town
- Identify how Apex officials carried out the post-fire investigation
- Explain how scene security issues were addressed during the post-fire cleanup

In the previous chapter, you were placed in the driver's seat alongside the Unified Command staff to experience the EQ Fire from their perspective. From the initial reports of a large vapor cloud through the first explosion to the evacuation of 17,000 citizens, you were there. The incident was very dynamic and posed many challenges to the small town of Apex, and you saw how resources from many local, state, and federal agencies came together to bring the incident to a successful end. However, as with any incident that you may respond to, just because the fire is out does not mean we get to go home. This was the case after the EQ Fire. After the fire was out and because this incident drew so much national media attention, the questions began: What happened? Is the air safe? Can our children go outside to play? And these questions needed answers. In this chapter, we will discuss the post-fire investigation, cleanup activities, and environmental testing that had to be done to ensure everything was in fact "back to normal" for the citizens of Apex.

Reentry after the Event

After the fires at the Environmental Quality facility in Apex were out and the chemicals were contained, the city had to be repopulated. But first, we had to ask ourselves as first responders, Is this area safe? A panel of professionals aided Command Staff in making this determination. The panel consisted of Toxicologists from the EPA, Center for Toxicology Environmental Health (CTEH), Wake County Public Health, North Carolina DENR, and the North Carolina Office of Public Health. Determining the area's safety was no easy task. The panel had to review air samples taken from the site as well as around the perimeter. This was done through mobile air monitoring vehicles as well as through wireless air monitoring devices placed throughout the perimeter to check the air for contamination (see Figures 9-1 and 9-2).

Throughout the event, there were over 300,000 air samples taken. The air quality group reviewed the data and finally determined that the air was, in fact, safe for reentry. Up to this point, the Fire Chief had served as the Lead Spokesperson for the Unified Command Staff since the event up to this point had been primarily a fire event. It was now decided that this role should transition to the Police Chief for repopulation. The Unified

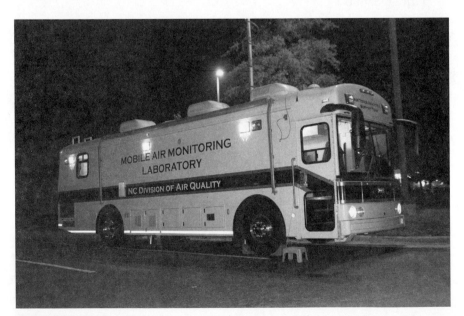

Figure 9-1 *North Carolina Division of Air Quality's mobile air monitoring unit is set up to test air samples from the area around the fire scene. Courtesy of Lee Wilson, photographer.*

Figure 9-2   *A wireless air monitoring device attached to a pole. Courtesy of the Town of Apex.*

Command Staff agreed on this transition since effective crowd and traffic control is primarily a law enforcement function. Also, the Fire Chief needed the time to be involved in the decision-making processes of the on-going post-fire investigation. The Fire Chief and EMS Chief remained as part of the Unified Command Staff to ensure that the fire scene was maintained and safety was still paramount. The Police Chief developed a phased reentry plan that used major roads and subdivision lines to separate the affected portions of the town into manageable quadrants. See Figure 9-3 for a map that was produced of the evacuated area. The reentry phases were identified by number and color to alert citizens when their neighborhood would be opened. The map was used by Public Information Staff in Apex and posted on all local media.

Police officers were positioned at strategic points along the reentry routes. With control points in place to make sure that all reentry traffic was maintained on identified travel routes, Apex was reopened for business, and residents were allowed to return home. These same control points were also used to verify that people reentering the town were supposed to be there: citizens along with business and property owners. An event this large draws a lot of curious onlookers.

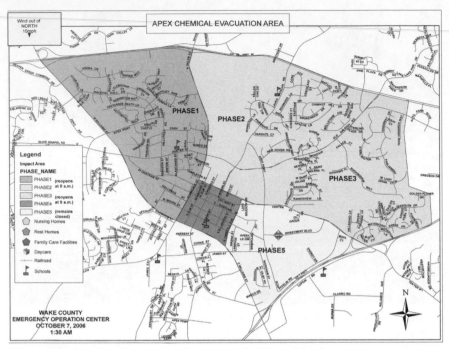

Figure 9-3   *Reentry Phases Map. Courtesy of the Town of Apex.*

Public Information Officials released information via the media to returning residents regarding safety and what to be aware of as they returned
home. The last area to be repopulated was the area around the EQ facility.
A three-block area around the facility site remained off-limits to the general
public and to neighboring businesses for another three days until a site
perimeter could be established. Business owners were allowed to check their
businesses for damage but not reopen them until the area could be determined safe for occupancy. Law enforcement officers were staged around the
site perimeter to secure the facility against entry until the post-fire investigation could be completed. Intrigued media and citizens wanted to get closer
to the area of the fire. For this reason, police officers were left in place at the
EQ facility for almost six weeks post-event, providing a security presence
initially for the fire investigation team and then the cleanup crews who arrived to dispose of the large quantities of hazardous waste left after the fire.

The Investigation

While the Police Chief and law enforcement officials were taking care of
the repopulation of Apex, fire officials were preparing to enter the site to

perform a cause and origin investigation. The investigative team was made up of representatives from the Wake County Fire Marshal's Office; Alcohol, Tobacco and Firearms; and North Carolina State Bureau of Investigation. This group was responsible for identifying the cause and origin of the fire. The fire investigators had to deal with many unusual situations while investigating this fire scene. Each member of the team had to be outfitted in full protective gear to include firefighting clothing, chemical protective suits, and breathing apparatus. The fact that the protective ensemble was hot and very bulky and the breathing apparatus only provided 30 minutes of air limited the team as to how long they could operate in the fire scene. The team also had to be very cautious of what they disturbed while searching through the scene of the fire. Even though the fires had been extinguished, there were still chemical contamination concerns and many containers that had not yet ruptured but were pressurized due to the heat impingement from the fire. This created some potential extreme hazards for our fire investigators. In Figures 9-4 and 9-5, members of the investigative team sift through the facility remains accompanied by a chemical toxicologist to ensure safety.

Figure 9-4 *Members of the Investigative Team sift through the rubble. Courtesy of the Wake County Fire Marshal.*

Figure 9-5 *Investigation hazards. In the background are containers. A chemical specialist accompanied each entry team from the investigative group to ensure their safety and monitor the air around the site. Courtesy of the Wake County Fire Marshal.*

These figures show the numerous opened chemical containers that investigators had to sift through. With the warehouse structure collapsed, there was a large amount of metal debris and equipment that also created a hazardous environment for the investigators.

While the on-scene investigation was going on, there was another group from the North Carolina State Bureau of Investigation interviewing EQ employees and other witnesses, including first responders, in an effort to better understand the events surrounding this fire and explosion.

For the investigators who had entered the scene and sifted through the warehouse site, a decontamination area was set up, and all personnel were decontaminated as they exited the area. All equipment that was worn or used while in the fire scene had to be decontaminated as well (see Figure 9-6).

The Site Cleanup

After the investigative team completed their site survey and it had been determined that the scene was secure, the Fire Chief turned the property over

Figure 9-6   *All investigators and equipment were decontaminated. Courtesy of the Wake County Fire Marshal.*

to the environmental cleanup companies that EQ had hired. The companies were monitored by both the EPA and North Carolina Department of Environmental and Natural Resources to ensure that the cleanup and removal of all potential contaminants was completed in an effective, safe, and efficient manner. The Apex Police Department maintained site security through the cleanup process for the cleanup contractors and to make sure that no unauthorized persons entered the secure area. This cleanup process lasted for six weeks. During that time, all materials from the building were removed, packaged, and hauled away as were the remains of the chemical containers. In Figure 9-7, you can see the large amounts of chemical containers and other equipment that had to be identified, packaged, and removed by the cleanup contractors.

The local first responders worked very closely with the contracted cleanup companies. Even though the initial removal of debris from the site was completed in six weeks, contractors were required to remain on-site for several more weeks, cleaning the area and performing sample tests to ensure that no environmental hazards remained. In the Apex event, EQ brought in two contractor groups, which included an industrial firefighting

Figure 9-7   *Materials needing to be collected, packaged, and removed from the site. Courtesy of the Wake County Fire Marshal.*

team and cleanup company as well as an environmental engineering firm specializing in chemical toxicology. Both groups worked closely with local first responders and fire hazardous materials teams throughout the cleanup efforts. During the final stages of the cleanup, contractors brought in heavy equipment to complete the debris removal process (see Figure 9-8).

Post-Event Media

During the fire, the media was a beneficial resource to Apex officials. They assisted with public notification and, through the Apex Public Information Officer, shared necessary information with the community. After the fire was extinguished and the town repopulated, the media still had to do their job—report the news. In this case, the news was the fallout from such a large chemical fire and how the community had been affected. The media was at the EQ site every day following up on the progress of the cleanup efforts, inquiring from neighboring businesses how the fire affected them and the citizens of Apex. Town officials had to ensure that the media questions were answered and at the same time address the many questions coming directly from the community. The Apex Public Information Officer

Figure 9-8   *On-site debris cleanup. Courtesy of the Apex Fire Department.*

used the media to address their questions. Initially, press conferences were held several times each week to answer questions about the cleanup process and reassure the community that all was safe in Apex (see Figure 9-9). Apex hired an Environmental Contractor to verify that the air, water, and facilities around the site were in fact clean and safe from any potential contamination. Apex officials wanted to bring in an independent environmental toxicology group to perform samples of the air, water, and ground in the affected area around the site. This was done to ensure citizens that the town had made every effort to address concerns of long-term environmental damage from the fire. By utilizing the news media to communicate daily activities from the EQ site to the citizens, the town officials could keep citizens informed of the ongoing activities to bring life back to some sense of normalcy.

The EQ Event Goes Out with a Bang

With the fires extinguished and the scene secure, town officials thought the worst was behind them. Then on October 18, 2006, the fire department received a call for an explosion at the EQ facility where the cleanup was in

Figure 9-9   *Apex Mayor Keith Weatherly during one of many news interviews to discuss cleanup efforts at the EQ site. Courtesy of the Town of Apex.*

progress. Immediately, on-scene police officers confirmed that there had been an explosion and that there now appeared to be a fire in progress in the area. It seemed to be starting all over. Ironically enough, the same shift that was working on the night of the first fire was also on duty on October 18. For the on-duty Battalion Chief, it was déjà vu. As units were responding from the Apex Fire Department, they were being given updates on what had occurred. As the first fire unit arrived, dispatch advised that the explosion had originated from a 55-gallon drum of sodium metal that had gotten wet and reacted violently to the moisture. On-scene firefighters worked with hazmat specialists and the cleanup contractors to secure the situation (see Figure 9-10). In this case, the material was allowed to burn out and then it was overpacked in a transportation container to be removed from the site. After this second fire event, contractors were required to go back and revisit their cleanup plan to address the possibility of any remaining products that could react in a similar manner.

Since the EQ Fire and the subsequent blast during the cleanup operations, life has returned to normal in this small southern town. But town officials and first responders know that even in a small town, major incidents can occur. They also know that by being prepared, you can respond to these events in a manner that will protect the community. By planning for such an event, officials in Apex were able to keep citizens safe and maintain

Figure 9-10 *First responders discuss a second explosion during cleanup. Courtesy of the Town of Apex.*

continuity of government services even in the face of adverse conditions caused by this fire. There were many lessons learned from this event, and town officials applied those lessons in an effort to enhance their plan. Below are some of those efforts:

- Develop a Continuity of Operations Program as part of the current EOP to address alternative facilities for continued governmental operations.
- Ensure that management staff have the appropriate relief and rehabilitation. During the Apex incident, operational personnel were relieved regularly, but management staff worked extended hours with little relief.
- Implement software to aid in event tracking and data collection during the incident.
- Make sure all personnel have the appropriate personal protective equipment. This can be addressed by providing police officers with equipment to provide exposure protection from chemicals.
- Provide primary mover vehicles to tow equipment trailers. Both the fire and police departments in Apex had equipment trailers but were unable to move all the equipment due to a limited number of towing vehicles.
- Ensure all necessary mutual aid agreements are in place.

Chapter Summary

This chapter addressed the many issues that arose after the EQ Fire was out. Of course, just because the fire is out, the job is not over—by far. After the EQ Fire, officials were still faced with the post-event investigation and answering the many questions that both the community and the media had. Apex officials also had to make sure that the site remained secure from intrusion during the cleanup activities. To ensure that the areas around the site were safe from contamination, local officials hired their own environmental firm to take samples of all affected areas and then verify that the samples were free from contamination. This chapter identifies the lessons learned post-event and how Apex used these lessons to enhance its local plans.

Review Questions

1. What were some of the challenges faced by fire investigators as they performed their investigations at the fire scene?
2. How did Apex keep citizens informed after the flurry of activities occurring at the EQ facility?
3. How did town officials verify that certain areas of the town were free from chemical contamination after the fire?

Acronyms

CTEH—Center for Toxicology and Environmental Health

DENR—Department of Environmental and Natural Resources

EOC—Emergency Operations Center

EOP—Emergency Operations Plan

EPCRA—Emergency Planning and Community Right-to-Know Act

HSOC—Homeland Security Operations Center

IAP—Incident Action Plan

ICS—Incident Command System

IMT—Incident Management Team

JFO—Joint Field Operations

JIC—Joint Information Center

JOC—Joint Operations Center

LEPC—Local Emergency Planning Committee

MAC—Multi-Agency Coordination

MSDS—Material Safety Data Sheets

NCCERT—North Carolina Canine Emergency Response Team

NIMS—National Incident Management System

NRCC—National Response Coordination Center

PIO—Public Information Officer

RCRA—Resource Conservation and Recovery Act

RRCC—Regional Response Coordination Center

SOG—Standard Operating Guidelines

Glossary

All Hazard Plan–the concept of an "All Hazard Plan" is one in which a target hazard assessment is completed and a plan to address identified hazards is then written taking into consideration all aspects of preparation, response, mitigation, and recovery.

Area Command–is an expansion of the incident command function primarily designed to manage a very large incident or area that has multiple incident management teams assigned. The function is to develop broad objectives for the impacted area and coordinate the development of individual incident objectives and strategies.

Area Commander (Single or Unified Area Command)–responsible for the overall direction of incident management teams assigned to the same incident or to incidents in close proximity. The responsibility includes ensuring that conflicts are resolved, compatible incident objectives are established, and strategies are selected for the use of critical resources.

Branch–that organizational level having functional, geographical, or jurisdictional responsibility for major parts of the incident operations. The Branch level is organizationally between Section and Division/Group in the Operations Section, and between Section and Units in the Logistics Section. Branches are identified by the use of Roman numerals, by function, or by jurisdictional name.

Branch Tactical Planning–method of expanding the planning capability by allowing detailed action plans to be developed within the Operations Section at the Branch level with the Planning Section providing support and coordination.

Chain of Command–chain of command refers to the orderly line of authority within the ranks of the incident management organization.

Command–the lead role in the Incident Command System. Command is responsible for every action taken on the scene and every decision made unless the responsibility is assigned to another individual.

Command Vehicle–any vehicle designated as the Incident Command Post. Generally, a vehicle specifically designed to serve as a Command Post with necessary resources such as communications equipment, area maps, and work stations but may simply be a chief officer's vehicle.

Continuity of Government–concept of ensuring that local government business can go forward after a disaster. Continuity of Government Plan should describe mitigation strategies and alternative methods to ensure that core business functions will continue.

Control Point–identified point generally along a roadway or at an intersection where traffic and/or people are routed or kept from entering.

Division–the organizational level responsible for operations within a defined geographic area. The Division level is organizationally between the Strike Team and the Branch.

Emergency Management Assistance Compact–organization endorsed by Congress in 1996, that is made up of all 50 states, the District of Columbia, Puerto Rico, and the Virgin Islands that serves as an interstate disaster assistance program. EMAC serves as a mutual aid agreement between participating states that uses standardized forms for assistance requests and allows affected states to request resources from other participating states that are willing to donate them.

Emergency Operations Center (EOC)– location used to coordinate information and resources to support local incident management activities.

Emergency Operations Plan (EOP)– plans used to identify responsibilities during an event. Organizational management structure, event operations, responsibilities, and incident objectives are generally defined in this plan.

Emergency Planning Zone (EPZ)– predetermined area surrounding an identified potential high hazard site or facility where emergency planning activities are implemented to include community notification and evacuation. As an example, the Town of Apex is in the EPZ for the Harris Nuclear Facility.

Evacuation Plan–plan designed to aid in the coordinated movement of people from a defined area. Generally these plans will identify preferred routes of travel and traffic redirection plans to aid in the orderly movement and relocation of citizens.

Exclusionary Zone–another name for the Hot Zone, this area is identified as the high hazard area and thus cordoned off to the general public.

Group–groups are established to divide the incident into functional areas of operation. Groups are located between Branches (when activated) and Resources in the Operations Sections.

Hazardous Materials Branch–an organizational level having functional, geographical, or jurisdictional responsibility for major parts of the incident operation. The Hazardous Materials Branch would be charged with responsibility of hazardous materials involved, such as in the EQ event.

Helibase–a location within the general incident area for parking, fueling, maintenance, and loading of helicopters.

Incident Action Plan (IAP)–your game plan or business plan for the event. It should identify what you want to do and who is responsible for doing it; how it will be done; how you will communicate with resources; and what the Medical Plan is for on-scene personnel. It may be written or verbal depending on the nature and size of the event.

Incident Command System (ICS)– the Incident Command System is a standardized management tool for meeting the demands of small or large emergency or nonemergency situations. It represents "best practices" and has become the standard for emergency management across the country. As stated by NIMS: "The ICS is a management system designed to enable effective and efficient domestic incident management by integrating a combination of facilities, equipment, personnel, procedures, and communications operating within a common organizational structure, designed to enable effective and efficient domestic incident management. A basic premise of ICS is that it is widely applicable. It is used to organize both near-term and long-term field-level operations for a broad spectrum of emergencies, from small to complex incidents, both natural and man-made. ICS is used by all levels of government—Federal, State, local and tribal—as well as by many private-sector and nongovernmental organizations. ICS is also applicable across disciplines. It is normally structured to facilitate activities in five major functional areas: command, operations, planning, logistics and finance and administration."

Incident Commander–the lead role responsible for organization and operational management. Provides overall incident objectives and strategies, establishes procedures for incident resource ordering, and establishes procedures for resource activation, mobilization, and employment. The Incident Commander approves the completed IAP with his or her signature. With the Safety Officer, he or she reviews hazards associated with the incident and proposed tactical assignments, assists in developing safe tactics, and develops the safety message.

Incident Complex–incident organizational option that allows for the addition of a second Logistics or Operations Section to overcome management challenges.

Information/Intelligence Function–a function within the organizational management structure that is responsible for developing, conducting, and managing information related to security plans and operations as directed by the Incident Plan. Can be a part of the Command Staff, a Unit within the Planning Section, a Branch within the Operations Section or a separate Section of the General Staff.

Incident Management–process of ensuring that emergency assessment, hazard operations, and population protection are undertaken in a timely manner and that responders have sufficient resources to do their jobs.

Incident of National Significance–an incident where resources from state and local authorities are overwhelmed, situations where more than one federal agency is involved with terrorist threats or high-profile, large-scale planned events or situations where the Department of Homeland Security is requested to assist under its own authority.

Incident Stabilization–the act of bringing a situation under control.

Increased Readiness–the act of placing all identified resources in an alert status ensuring that those resources are ready to respond if needed.

Life Safety–anything that may threaten the welfare of human life.

Local Emergency Planning Committee (LEPC)–a group made up from local community leaders, business professionals, and emergency responders responsible for the development and maintenance of planning data specifically targeting chemical emergencies.

Logistics–handles special transportation and communication needs and implements vehicle inspection program.

Major/Complex Incident–an incident that involves more than one agency or political jurisdiction, involves complex management and communications issues, requires numerous tactical and support resources, generally involves widespread damage and numerous injuries, lasts for multiple operational periods, and draws national media attention.

Manifest–written document listing an inventory.

Medical Branch–organizational level having functional, geographical, or jurisdictional responsibility. The Medical Branch is charged with overseeing on-scene medical operations.

Mitigation (functional area of activity)–an activity that actually eliminates or reduces the probability of a disaster occurrence, or reduces the effects of a disaster. Mitigation includes such actions as zoning and land use management, safety and building codes, flood proofing of buildings, and public education.

Multi-Point Ordering–ordering of resources from several different ordering points and/or the private sector.

Mutual Aid Agreement–formal or informal understandings between jurisdictions or organizations that pledge exchange of emergency or disaster assistance.

National Significance (per NIMS)–a system developed by the Department of Homeland Security designed to standardize disaster response. This system was developed based on Presidential Homeland Security Directives 5 and 8 addressing command and management, preparedness, resource management, communications and information management, support technologies, and maintenance.

Operational Period–time-defined period that operational objectives are assigned during an incident.

Operations Section–identifies continuing needs for operational resources and those that are, or will be, excess to the incident, and prepares the list for the Demobilization Unit Leader.

Pet Plan–a part of the EOP that addresses the needs of pets and animals involved in the incident.

Planning Meeting–meeting held by the Planning Section Chief to review and validate the operational plan and identify resource requirements.

Planning P–planning tool used to guide planning officials through the planning process.

Planning Section Chief–individual responsible for providing planning services to the incident.

Pre-Event Planning–the act of identifying potential problems, concerns, and resource needs associated with an event and developing plans to address those areas identified prior to the actual event.

Pre-Impact–actions accomplished that are precautionary, centered around taking appropriate measures to protect people, such as relocation, shelter inspections, facility security, and equipment readiness.

Pre-Incident Planning–the act of identifying potential hazards or threats within a specific jurisdiction and developing response plans to address those concerns.

Preparedness (functional area of activity)–ongoing process of continuous planning for a community to ensure that the emergency plan remains consistent to address identified needs based on potential threats.

Property Preservation–actions directed toward the mitigation of specific threats or hazards to preserve property.

Recovery (functional area of activity)–activities that involve assistance to enhance the return of the community to normal or near-normal conditions. Short-term recovery returns vital life-support systems to minimum operating standards. Long-term recovery may continue for a number of years after a disaster and seeks to return life to normal or improved levels. Recovery activities include temporary housing, loans or grants, disaster unemployment insurance, reconstruction, and counseling programs.

Resource Sufficiency Assessment–looking at your response capabilities compared to what each area of concern will require for an effective response.

Resource Unit Leader–the person responsible for maintaining the status of all assigned resources (primary and support) at an incident. Overseeing the check-in of all resources, maintaining a status-keeping system indicating current location and status of all resources and maintenance of a master list of all resources.

Response (functional area of activity)–activities occurring immediately before, during, and directly after an emergency or disaster. They involve lifesaving actions such as the activation of warning systems, manning the EOCs, implementation of shelter or evacuation plans, and search and rescue.

Reverse 911 System–a type of communication system that uses existing 911 system data such as phone numbers for a jurisdiction and dials out to the numbers in the system to deliver emergency messages for public notification.

Safety Analysis–tool used by the Safety Officer as a concise way of identifying hazards and risks present in different areas of the incident. Used to identify specific ways of mitigating these issues and concerns.

Safety Officer–the person responsible for developing and recommending measures to assure personnel safety, and to assess and/or anticipate hazardous and unsafe situations. Having full authority of the Incident Commander, the Safety Officer can exercise emergency authority to stop or prevent unsafe acts.

Safety Plan–written document within the IAP that identifies specific risks to operational resources and the identified safety/mitigation measures.

Section–the organizational level having functional responsibility for primary segments of incident management (Operations, Planning, Logistics, Finance/ Administration). The Section Level is organizationally between Branch and Incident Commander.

Sheltering in Place–a means of protecting individuals in their own homes or place of refuge when an event has already affected the area they are in. During shelter in place operations, the individual must close all windows and doors, turn off outside vent systems, and secure their place of refuge.

Sheltering Plan–generally part of the EOP or can serve as a stand-alone plan used to identify shelter sites and gives direction as to the management of sheltering sites.

Single-Point Ordering–resource ordering process where all resources are ordered through a single dispatch/ ordering center.

Single Resources–an individual piece of equipment and its personnel complement, or an established crew or team of individuals with an identified work supervisor that can be used on an incident.

Span of Control–an ICS concept that describes the ratio of individuals supervised to the number of supervisors. Under NIMS, an appropriate span of control is a ratio between 3:1 and 7:1 (between three and seven individuals supervised to one supervisor).

Staging Area–a preselected location having large parking areas such as a major shopping area or school. The area is a base for the assembly of persons to be moved by public transportation to host jurisdictions and a debarking area for returning evacuees. Several of these areas should be designated to each evacuating jurisdiction.

Standard Operating Guidelines (SOG)–a set of instructions covering those features of operations that lend themselves to a guidance without loss of effectiveness.

Strike Team–specific combination of similar kind and type of resources that have common communications and a leader.

Systematic Route Alerting–method of community notification where emergency vehicles travel predetermined routes announcing necessary emergency information to the public.

Tactics Meeting–part of the planning process where tactics are developed by the Operations Section Chief.

Task Force–group of resources with common communications and a leader that may be preestablished and sent to an incident; or they can be formed during the event.

Unified Command–a team approach that allows all agencies with jurisdictional responsibility for an incident, either geographical or functional, to participate in the management of the incident.

Unity of Command–concept that personnel report only to one supervisor and receive assignments from only their supervisor.

Defining NIMS Typing

A. Purpose

This appendix provides additional information regarding the national equipment typing system specified in Chapter 4 of this document.

B. Responsibilities

The NIMS Integration Center described in Chapter 7 has the overall responsibility for ongoing development and refinement of various NIMS activities and programs. Under its auspices, the National Resource Management Working Group, chaired by the Emergency Preparedness and Response Directorate of the Department of Homeland Security, is responsible for establishing a national resource typing protocol. The NIMS resource typing protocol is based on inputs from representatives from various Federal agencies and departments and private organizations, as well as representatives of State and local emergency management; law enforcement; firefighting and emergency medical services; public health; public works; and other entities with assigned responsibilities under the Federal Response Plan and the National Response Plan. Federal, State, local, and tribal authorities should use the national typing protocol when inventorying and managing resources to promote common interoperability and integration.

C. Elements of the National Typing Protocol

The resource typing protocol provided by the NIMS describes resources using category, kind, components, metrics, and type data. The following data definitions will be used:

1. RESOURCE
   For purposes of typing, resources consist of personnel, teams, facilities, supplies, and major items of equipment available for

assignment or use during incidents. Such resources may be used in tactical support or supervisory capacities at an incident site or EOC. Their descriptions include category, kind, components, metrics, and type.

2. CATEGORY

A category is the function for which a resource would be most useful. Table A-1 briefly describes the categories used in the national resource typing protocol.

Category	Purpose
Transportation	To assist Federal agencies, State and local governments, and voluntary organizations requiring transportation to perform incident management missions following a major disaster or emergency; to coordinate incident management operations and restoration of the transportation infrastructure.
Communications	To provide communications support for Federal, State, local, and tribal incident management efforts.
Public works and engineering	To assist those engaged in life saving, life-sustaining, damage mitigation, and recovery operations following a major disaster or emergency by providing technical advice, evaluation, and engineering services; by contracting for construction management and inspection and for the emergency repair of water and wastewater treatment facilities; supplying potable water and ice and emergency power; and arranging for needed real estate.
Firefighting	To detect and suppress urban, suburban, and rural fires.

Table A-1   *Categories Used in the National Resource Typing System.*

Category	Purpose
Information and planning	To collect, analyze, process, and disseminate information about a potential or actual disaster or emergency to facilitate overall activities in providing assistance to support planning and decision-making.
Law enforcement and security	To provide law enforcement assistance during response and recovery operations; to assist with site security and investigation.
Mass care	To support efforts to meet the mass care needs of disaster victims including delivering such services as supplying victims with shelter, feeding, and emergency first aid; supplying bulk distribution of emergency relief supplies; and collecting information to and for a disaster welfare information system designed to report on victim status and assist in reuniting families.
Resource management	To provide operational assistance for incident management operations.
Health and medical	To provide assistance to supplement local resources in meeting public health and medical care needs following a disaster or emergency or during a potential developing medical situation.
Search and rescue	To provide specialized lifesaving assistance in the event of a disaster or emergency, including locating, extricating, and providing on-site medical treatment to victims trapped in collapsed structures.
Hazardous materials response	To support the response to an actual or potential discharge and/or release of hazardous materials.

Table A-1 *Continued*

Category	Purpose
Food and water	To identify, secure, and arrange for the transportation of safe food and water to affected areas during a disaster or emergency.
Energy	To help restore energy systems following a disaster or emergency.
Public information	To contribute to the well-being of the community following a disaster by disseminating accurate, consistent, timely, and easy-to-understand information; to gather and disseminate information about disaster response and recovery process.
Animals and agricultural issues	To coordinate activities responding to an agricultural disaster and/or when the health or care of animals is at issue.
Volunteers and donations	To support the management of unsolicited goods and unaffiliated volunteers, and to help establish a system for managing and controlling donated goods and services.

Table A-1　*Continued*

3. KIND

Kind refers to broad classes that characterize like resources, such as teams, personnel, equipment, supplies, vehicles, and aircraft.

4. COMPONENTS

Resources can comprise multiple components. For example, an engine company may be listed as having the eight components shown in Table A-2.

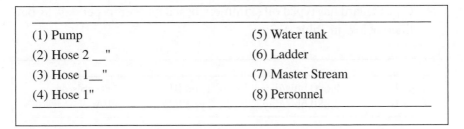

(1) Pump	(5) Water tank
(2) Hose 2 __"	(6) Ladder
(3) Hose 1__"	(7) Master Stream
(4) Hose 1"	(8) Personnel

Table A-2 *Example of a Resource with Multiple Components (Fire Fighting Engine Company).*

As another example, urban search and rescue (US&R) teams consist of two 31-person teams, four canines, and a comprehensive equipment cache. The cache is divided into five separate, color-coded elements and is stored in containers that meet specific requirements.

5. METRICS

Metrics are measurement standards. The metrics used will differ depending on the kind of resource being typed. The mission envisioned determines the specific metric selected. The metric must be useful in describing a resource's capability to support the mission. As an example, one metric for a disaster medical assistance team is the number of patients it can care for per day. Likewise, an appropriate metric for a hose might be the number of gallons of water per hour that can flow through it. Metrics should identify capability and/or capacity.

6. TYPE

Type refers to the level of resource capability. Assigning the Type I label to a resource implies that it has a greater level of capability than a Type II of the same resource (for example, due to its power, size, or capacity), and so on to Type IV. Typing provides managers with additional information to aid the selection and best use of resources. In some cases, a resource may have less than or more than four types; in such cases, either additional types will be identified, or the type will be described as "not applicable." The type assigned to a resource or a component is based on a minimum level of capability described by the identified metric(s) for that resource. For example, the U.S.

Coast Guard has typed oil skimmers based on barrels per day, as outlined below in Table A-3:

| Type I | 9,600 bbls/day | Type III | 480 bbls/day |
| Type II | 2,880 bbls/day | Type IV | N/A |

Table A-3

7. ADDITIONAL INFORMATION

The national resource typing protocol will also provide the capability to use additional information that is pertinent to resource decision-making. For example, if a particular set of resources can only be released to support an incident under particular authorities or laws, the protocol should provide the ability for resource managers to understand such limitations.

D. Example of a Resource for Which Typing Has Been Completed

As an illustration of how the national equipment typing system is used, Table A-4 is an example of a resource that has been completely typed, an urban search and rescue task force.

RESOURCE: US&R TASK FORCES

Category: Search & Rescue (ESF 9) Kind: Team

Minimum Capabilities (Component)	Minimum Capabilities (Metric)	Type I	Type II	Type III	Type IV	Other
Personnel	Number of People per Response	70-person response.	28-person response.			
Personnel	Training	NFPA 1670 Technician Level in area of specialty. Support personnel at Operations Level.	NFPA 1670 Technician Level in area of specialty. Support personnel at Operations Level.			
Personnel	Areas of Specialization	High angle rope rescue (including highline systems); confined space rescue (permit required); Advanced Life Support (ALS) intervention; communications; WMD/HM operations; defensive water rescue.	Light frame construction and basic rope rescue operations; ALS intervention; HazMat conditions; communications; and trench and excavation rescue.			
Personnel	Sustained Operations	24-hour S&R operations. Self-sufficient for first 72 hours.	12-hour S&R operations. Self-sufficient for first 72 hours.			

Table A-4

Continued

Minimum Capabilities (Component)	Minimum Capabilities (Metric)	Type I	Type II	Type III	Type IV	Other
Personnel	Organization	Multidisciplinary organization of Command, Search, Rescue, Medical, HazMat, Logistics, and Planning.	Multidisciplinary organization of Command, Search, Rescue, Medical, HazMat, Logistics, and Planning.			
Equipment	Sustained Operations	Potential mission duration of up to 10 days.	Potential mission duration of up to 10 days.			
Equipment	Rescue Equipment	Pneumatic Powered Tools, Electric Powered Tools, Hydraulic Powered Tools, Hand Tools, Electrical, Heavy Rigging, Technical Rope, Safety.	Pneumatic Powered Tools, Electric Powered Tools, Hydraulic Powered Tools, Hand Tools, Electrical, Heavy Rigging, Technical Rope, Safety.			
Equipment	Medical Equipment	Antibiotics/Antifungals, Patient Comfort Medication, Pain Medications, Sedatives/Anesthetics/Paralytics, Steroids, IV Fluids/Volume, Immunizations/Immune Globulin, Canine Treatment,	Antibiotics/Antifungals, Patient Comfort Medication, Pain Medications, Sedatives/Anesthetics/Paralytics, Steroids, IV Fluids/Volume, Immunizations/Immune Globulin, Canine Treatment,			

Equipment	Basic Airway, Intubation, Eye Care Supplies, IV Access/ Administration, Patient Assessment Care, Patient Immobilization/Extrication, Patient/PPE, Skeletal Care, Wound Care, Patient Monitoring.	Basic Airway, Intubation, Eye Care Supplies, IV Access/Administration, Patient Assessment Care, Patient Immobilization/Extrication, Patient/PPE, Skeletal Care, Wound Care, Patient Monitoring.
Technical Equipment	Structures Specialist Equip., Technical Information Specialist Equip., HazMat Specialist Equip., Technical Search Specialist Equip., Canine Search Specialist Equip.	Structures Specialist Equip., Technical Information Specialist Equip., HazMat Specialist Equip., Technical Search Specialist Equip., Canine Search Specialist Equip.
Communications Equipment	Portable Radios, Charging Units, Telecommunications, Repeaters, Accessories, Batteries, Power Sources, Small Tools, Computer.	Portable Radios, Charging Units, Telecommunications, Repeaters, Accessories, Batteries, Power Sources, Small Tools, Computer.
Equipment		

Table A-4

Continued

Minimum Capabilities (Component)	Minimum Capabilities (Metric)	Type I	Type II	Type III	Type IV	Other
Equipment	Logistics Equipment	Water/Fluids, Food, Shelter, Sanitation, Safety, Administrative Support, Personal Bag, Task Force Support, Cache Transportation/Support, Base of Operations, Equipment Maintenance.	Water/Fluids, Food, Shelter, Sanitation, Safety, Administrative Support, Personal Bag, Task Force Support, Cache Transportation/Support, Base of Operations, Equipment Maintenance.			

Comments:

Federal asset. There are 28 FEMA US&R Task Forces, totally self-sufficient for the first 72 hours of a deployment, spread throughout the continental United States trained and equipped by FEMA to conduct physical search and rescue in collapsed buildings, provide emergency medical care to trapped victims, assess and control gas, electrical services and hazardous materials, and evaluate and stabilize damaged structures.

FEMA

NIMS IMPLEMENTATION ACTIVITY SCHEDULE

The matrix below summarizes by Federal Fiscal Year (FY) all on-going NIMS implementation activities that have been prescribed by the NIMS Integration Center in FYs 2005 and 2006, as well as the **seven** new activities for States and territories and **six** new activities for tribes and local jurisdictions required in FY 2007. **State, territorial, tribal, and local jurisdictions should bear in mind that implementation activities from previous fiscal years remain on-going commitments in the present fiscal year. Jurisdictions must continue to support *all* implementation activities required or underway in order to achieve NIMS compliance.**

Future refinement of the NIMS will evolve as policy and technical issues are further developed and clarified. As a result, the NIMS Integration Center may issue additional requirements to delineate what constitutes NIMS compliance in FY 2008 and beyond. With the completion of the FY 2007 activities, State, territorial, tribal and local jurisdictions will have the foundational support for future NIMS implementation and compliance. The effective and consistent implementation of the NIMS nationwide will result in a strengthened national capability to prevent, prepare for, respond to and recover from any type of incident.

	NIMS IMPLEMENTATION ACTIVITY	FY 2007 STATE/TERRITORY	FY 2007 TRIBAL/LOCAL	FY 2006 STATE/TERRITORY	FY 2006 TRIBAL/LOCAL	FY 2005 STATE/TERRITORY	FY 2005 TRIBAL/LOCAL
ADOPTION	1. Support the successful adoption and implementation of the NIMS.					✓	✓
	2. Adopted NIMS for all government departments and agencies, as well as promote and encourage NIMS adoption by associations, utilities, non-governmental organizations (NGOs) and private sector incident management and response organizations.			✓	✓		
	3. Monitor formal adoption of NIMS by all tribal and local jurisdictions.			✓			
	4. Establish a planning process to ensure the communication and implementation of NIMS requirements, thereby providing a means for measuring progress and facilitate reporting.			✓			
	5. Designate a single point of contact to serve as the principal coordinator for NIMS implementation.		✓	✓			
	6. Designate a singe point of contact within each of the jurisdiction's Departments and Agencies.	✓					
	7. To the extent permissible by law, ensure that Federal preparedness funding, including DHS Homeland Security Grant Program and the Urban Areas Security Initiative (UASI), support NIMS implementation at the State and local levels and incorporate NIMS into existing training programs and exercises.					✓	
	8. To the extent permissible by law, ensure that federal preparedness funding to State and territorial agencies and tribal and local jurisdictions is linked to satisfactory progress in meeting FY2006 NIMS implementation requirements.			✓			
	9. To the extent permissible by State and territorial law and regulations, audit agencies and review organizations routinely include NIMS implementation requirements in all audits associated with federal preparedness grant funds, validating the self-certification process for NIMS compliance.			✓			
	10. Monitor and assess outreach and implementation of NIMS Requirements.	✓					
COMMAND AND MANAGEMENT	11. Coordinate and provide technical assistance to local entities regarding NIMS institutionalized use of ICS.					✓	
	12. Manage all emergency incidents and pre-planned (recurring/special) events in accordance with ICS organizational structures, doctrine and procedures, as defined in NIMS. ICS implementation must include the consistent application of Incident Action Planning and Common Communications Plans.			✓	✓		
	13. Coordinate and support emergency incident and event management through the development and use of integrated multi-agency coordination systems, i.e. develop and maintain connectivity capability between local Incident Command Posts (ICP), local 911 Centers, local Emergency Operations Centers (EOCs), the state EOC and regional and/ federal EOCs and NRP organizational elements.			✓	✓		
	14. Institutionalize, within the framework of ICS, the Public Information System (PIS), comprising of the Joint Information System (JIS) and a Joint Information Center (JIC).			✓	✓		
	15. Establish public information system to gather, verify, coordinate, and disseminate information during an incident.	✓	✓				
PREPAREDNESS: PLANNING	16. Establish NIMS baseline against the FY 2005 and FY 2006 implementation requirements.			✓	✓		
	17. Develop and implement a system to coordinate and leverage all federal preparedness funding to implement the NIMS.			✓	✓		
	18. Incorporate NIMS into Emergency Operations Plans (EOP).					✓	
	19. Revise and update plans and SOPs to incorporate NIMS and National Response Plan (NRP) components, principles and policies, to include planning, training, response, exercises, equipment, evaluation and corrective actions.			✓	✓		
	20. Promote intrastate mutual aid agreements, to include agreements with private sector and non-governmental organizations.					✓	✓
	21. Participate in and promote intrastate and interagency mutual aid agreements, to include agreements with the private sector and non-governmental organizations.			✓	✓		
PREPAREDNESS: TRAINING	22. Leverage training facilities to coordinate and deliver NIMS training requirements in conformance with the NIMS National Standard Curriculum.			✓			
	23. Complete training—IS-700 NIMS: An Introduction, IS-800 NRP: An Introduction; ICS-100 and ICS-200.			✓	✓		
	24. Complete training—ICS-300, ICS-400.	✓	✓				
PREPAREDNESS: EXERCISES	25. Incorporate NIMS/ICS into training and exercises.					✓	✓
	26. Participate in an all-hazard exercise program based on NIMS that involves responders from multiple disciplines and multiple jurisdictions.			✓	✓		
	27. Incorporate corrective actions into preparedness and response plans and procedures.			✓	✓		
RESOURCE MANAGEMENT	28. Inventory response assets to conform to FEMA Resource Typing standards.			✓	✓		
	29. Develop state plans for the receipt and distribution of resources as outlined in the National Response Plan (NRP) Catastrophic Incident Annex and Catastrophic Incident Supplement.			✓			
	30. To the extent permissible by state and local law, ensure that relevant national standards and guidance to achieve equipment, communication and data interoperability are incorporated into state and local acquisition programs.			✓	✓		
	31. Validate that inventory of response assets conforms to FEMA Resource Typing Standards.	✓	✓				
	32. Utilize response asset inventory for mutual aid requests, exercises, and actual events.	✓	✓				
COMMUNICATION & INFORMATION MANAGEMENT	33. Apply standardized and consistent terminology, including the establishment of plain language communications standards across public safety sector.			✓	✓		
	34. Develop systems and processes to ensure that incident managers at all levels share a common operating picture of an incident.	✓	✓				

October 2006

Town of Apex Fire Department Hurricane Policy and Procedures

Definitions

Condition 1—Increased readiness. Hurricane/tropical storm watch. See Section 11.

Condition 2—Pre-Impact. Hurricane/tropical storm warning. See Section 111.

Condition 3—Immediate Impact. Hurricane condition will affect the Town of Apex. See Section IV.

Condition 4—Sustained Emergency. Hurricane recovery phase. See Section V.

Hurricane Advisory—Official information issued by hurricane centers describing all tropical cyclone watches and warnings in effect along with details concerning tropical cyclone locations, intensity and movement, and precautions that should be taken. Advisories are also issued to describe (a) tropical cyclones prior to issuance of watches and warnings and (b) subtropical cyclones.

Hurricane Eye—The relatively calm area in the center of the storm. In this area, winds are light and the sky often is only partly covered by clouds.

Hurricane Season—That part of the year having a relatively high incidence of hurricanes. In the Atlantic, Caribbean, and Gulf of Mexico, the hurricane season is the period from June through November.

Hurricane Threat—The expectation that a Hurricane Watch will soon be announced. Hurricane Watch preparations may be initiated at this time by the Fire Chief.

Hurricane Warning—A warning that sustained winds of 64 knots (74 mph) or higher, associated with a hurricane, are expected in a specified coastal area within 24 hours or less. A hurricane warning can remain in effect when dangerously high water or a combination of dangerously high water and exceptionally high waves continue even though winds may be less than hurricane force.

Hurricane Watch—An announcement for specific areas that a hurricane or an incipient hurricane condition poses a possible threat to coastal areas generally within 36 hours.

Immediate Impact—Condition 3. Operations during a hurricane. Actions are concentrated on the well-being of people affected by the emergency. Emphasis is centered around life saving and property protection. Preliminary damage assessments begin.

Incident Action Plan—Initially prepared at the first planning meeting, the incident action plan contains general control objectives reflecting overall incident strategy, and specific action plans for the next operational period. When complete, incident action plans will have a number of attachments.

Increased Readiness—Condition 1. Hurricane preseason preparedness through hurricane threat/watch.

Normal Preparedness—Normal and special preparations and policy/procedure changes occurring all year long.

Operational Period—The period of time scheduled for execution of a given set of operation actions as specified in the IAP.

Post Emergency—Recovery efforts: Permanent restoration of private and public property. Return to normal service, repair and replace department assets, complete post-incident analysis, review/revise policies and procedures.

Pre-impact—Condition 2. Hurricane Warning or similar notification. Actions accomplished are precautionary, centered around taking appropriate measures to protect people, such as relocation, shelter inspections, facility security, etc.

Standby Status—The status of all Department employees upon the announcement of a Hurricane Watch. Requires all employees to prepare themselves and their homes and families for a possible hurricane landfall. Requires employees to monitor the status of the hurricane and be prepared to respond to the requirements of the Hurricane Policy and Procedure.

Sustained Emergency—Condition 4. Operations after a hurricane (Recovery Phase). Emphasis on helping injured and displaced persons and securing dangerous areas.

Town of Apex Fire Department Hurricane Policy and Procedures

Command and Control

I. PRE-HURRICANE SEASON

   A. All department command and Staff personnel will review and exercise Incident Command System throughout the year.

   B. The Chief of Fire Suppression will review this plan in May of each year for any changes and updates. Changes and updates to plan should be discussed in a staff meeting and then approved by the fire chief.

   C. All identified members of the hurricane command structure will attend preseason briefings scheduled by the Chief of Fire Suppression. These personnel will review the hurricane plan and all assignments and discuss any changes or additions to the plan.

   D. All personnel will update their individual family action plan and be informed of any changes and updates to the fire department's hurricane response plan.

II. CONDITION 1—INCREASED READINESS—HURRICANE WATCH

The National Hurricane Center has issued a hurricane watch for Southeastern North Carolina. All personnel must begin making preparations for a hurricane warning and begin taking steps to secure their family and property. The Chief of Fire Suppression must make all staff assignments.

   1. On duty personnel must begin making the following preparations at their respective stations.

      A. Check and police all grounds for any items or debris that could become flying projectiles during high wind conditions. All items not removed must be secured.

      B. Check all portable and station generators for adequate fuel.

      C. Locate any coolers or other containers for storing water and food.

   D. Fill all apparatus fuel tanks and determine the fuel supplies for all gasoline powered equipment and tools.

   E. Keep all portable radio batteries charged.

   F. Maintain a log of all expenses and alarms if a state of emergency is declared in addition to the daily log. See pre-printed forms.

   G. Off-duty personnel should begin making preparations for their family and home if a recall of personnel is needed.

   H. Check the station for adequate supplies that may be needed during and after the storm.

2. All town vehicles considered to be essential during or immediately following a hurricane will be readily available and completely fueled and serviced.

III. CONDITION 2—PRE-IMPACT—HURRICANE WARNING

The National Hurricane Center has issued a hurricane warning for Southeastern North Carolina. Hurricane conditions can be expected within 24 hours. All days off except regular will be cancelled. All off-duty personnel shall contact their respective station for information concerning a possible recall to duty. The Chief of Fire Suppression will institute this recall. Shift schedules may change if the need arises to meet the needs of the fire department.

A. All personnel reporting for duty whether for a regular shift or by recall should bring a supply of non-perishable food for a period of three days. Personnel will remain on duty until formally relieved.

B. Recheck the station and grounds for any unsecured items or debris.

C. Update information concerning the storm to all stations as conditions change.

D. All special requests for supplies (medical supplies, etc.) should be delivered ASAP.

E. Secure all loose items on each apparatus (hose cover, etc.).

F. Staff personnel assignments are as followed:

Wake County EOC

Town of Apex EOC

Fire Department Command Post

Personnel to be assigned by Fire Chief

Chief/Fire Marshal

House Captain

The Fire Department Command Post will be activated automatically when a Hurricane Warning is announced, if not previously activated by the Fire Chief.

911 Center

Designated representative

G. If the storm conditions are anticipated in Southeastern North Carolina before the on-duty personnel shift ends, all attempts will be made to allow personnel to return home and secure their residence. All personnel must implement their individual family action plan.

H. If shelters are open, the Fire Inspector will perform an inspection to insure all fire prevention codes are met.

 I. When a hurricane warning is announced, all burning permits will be cancelled. This prohibition will stay in effect until cancelled by the Chief of Fire Department.

IV. CONDITION 3—IMMEDIATE IMPACT—HURRICANE CONDITIONS

As conditions begin to deteriorate along the coast of Southeastern North Carolina, all preparations should have been completed. Operations during a hurricane: Actions are concentrated on the well-being of people affected by the emergency. Emphasis is centered around life saving and property protection. Preliminary damage assessments begin.

During such time as actual hurricane conditions exist, every attempt will be made to continue our primary mission of protecting lives and property in the Town of Apex. It should be remembered however, that fire department personnel are subject to the same environmental limitations as are members of the public.

A. Discontinuation of Response (NO RESPONSE)

1. The Fire Chief or his designee shall determine, in consultation with Fire Department Command Post, when the department will cease responding to calls due to the severity of the storm. This decision will then be announced by the Fire Department Command Post as a NO RESPONSE order. Prior to this announcement, any unit officer who feels that situations encountered are sufficiently dangerous to personnel at his/her location, should contact the Command Post. Unit officers who feel the need to continue operations past the announcement from Fire Department Command

Post must justify this decision through the Fire Department Command Post and receive permission to continue their current task.

2. The following guideline may be used to determine when apparatus should be placed in non-response mode during storm conditions:

a. **Suppression** unit operations may be terminated when sustained winds of greater than 60 mph exists or local conditions dictate unsafe conditions (localized flooding, downed wires, etc.).

B. Hurricane Eye Operations

Operations during the eye of the hurricane should concern themselves primarily with re-securing the fire station, if necessary, and assisting citizens who come to the fire station when it would be a danger to release them. All such activities during the hurricane eye shall be undertaken only if such operations can be completed in a safe manner. The safety of department personnel will remain the primary consideration during these operations. In all cases, Hurricane Eye Operations should be coordinated through the Fire Department Command Requests for assistance received by the 911 Center, which occur during unsafe conditions and when emergency units are not able to respond, will be prioritized and remain on a waiting list at the 911 Center for post hurricane assignment.

V. CONDITION 4—SUSTAINED EMERGENCY—RECOVERY PHASE

A. Operations after a hurricane. Emphasis is on helping injured and displaced persons and securing dangerous areas.

1. General Instructions

Work Schedule Information

a. Off-duty employees with pre-assigned responsibilities will assume those duties.

b. Other off-duty employees will return to their regular work schedule and location upon the National Hurricane Center announcement that hurricane warnings have been lowered. Employees unable to reach their normal workplace will notify their immediate supervisor.

c. All personnel not scheduled to report to duty immediately following a storm must follow the call-in procedure in Addendum A.

d. Anyone unable to report for duty must, if possible, follow the call-in procedure in Addendum A.

B. Assessments—One of the most important functions for emergency service personnel following a disaster is the need to evaluate the impact that the disaster has had upon departmental resources and jurisdictional responsibilities. This assessment may include observations of structural damages, flooding, injuries, (both to fire department personnel and citizens), access, fire load, water supply, status of critical resources (such as hospitals, power stations, etc.) status of transportation capabilities with regard to both road accessibility and the operational capability of fire department equipment. Wherever possible, this evaluation will be accomplished by nonresponding personnel.

C. Response Operations

1. Resuming Operations (RESUME RESPONSE)

   The Fire Chief or his designee shall make a determination, in consultation with Fire Department Command Post when the department can resume response operations. This decision will then be announced by the Fire Department Command Post as a RESUME RESPONSE order. Unit officers who believe it is safe to resume operations prior to this announcement shall contact Fire Department Command Post and state the conditions at their location and their need to begin operations. They will be authorized to respond only upon approval from the Fire Department Command Post (this approval will be authorized by the Fire Chief). If unable to contact Fire Department Command Post, the decision to approve such operations will rest with the Unit Officer. If unable to contact the Unit Officer, the decision will be the responsibility of the Senior Firefighter. Activities shall be undertaken only if such operations can be completed in a safe manner.

2. Post-Hurricane Station Roll Call and Disaster Assessment "Snapshot"

   A Station Roll Call will be conducted by the Fire Department Command Post as soon as weather conditions have subsided. A report to the Fire Department Command Post will be made utilizing the Disaster Assessment Snapshot report (see Assessment Annex). The report will include the status of all personnel and equipment at the station as well as the status of the facility. The condition of the surrounding neighborhood as well as standing water levels and visible access will also be reported.

E. Communications
   1. In the event the 911 Center loses communication capabilities, Apex Fire Dept. will assume emergency communication procedures based on existing SOG.
   2. Under emergency conditions during a disaster, all radio communications must provide only essential information.
   3. Apex Fire Dept. will operate on Channel 2 as directed by the Fire Chief
      a. Clear Text Communications
          Upon the resumption of response operations following a hurricane, all units will communicate in clear text (plain English). This is to eliminate all code signals that may not be understood by other jurisdictions. This step is taken in anticipation of mutual aid departments coming to the assistance of the Town of Apex. All incoming mutual aid units will also be requested to speak in clear text.
      b. Emergency Response
          (1) Dispatch through the 911 Center
               The primary response method, unless otherwise indicated, will be dispatched by the 911 Center. Units responding to requests from the Fire Alarm Office must advise that office of any problems encountered during dispatch, or any changes of assignment necessitated by personal observations.
          (2) Dispatch via the Fire Department Command Post
               Dispatch of units may be controlled by the Fire Department Command Post. This dispatch may be relayed from Fire Alarm, may be based upon information at the Fire Department Command Post, or may be a combination of both. The decision to utilize Fire Department Command Post dispatch will be coordinated with the 911 Center.
          (3) Self-dispatch based upon assessment
               It may be necessary for units to dispatch themselves due to lack of communications with other stations or the 911 Center. This should be done based upon the OIC's assessment of the situation at the time. Consideration must be given to performing additional assessment objectives similar to triage during multi-casualty incidents. With the exception of providing life-saving assistance, a search

will be conducted as soon as possible of predetermined priority areas (i.e., shelters, hospitals, mobile home parks).

(4) Primary search and rescue

Primary search and rescue may be implemented to support emergency response. This would be limited to rescue of lightly trapped victims. Appropriate decisions must be made as to the priority of responsibilities during this time period.

(5) Safety & Hazard Identification

Personnel conducting emergency operations must realize that their own safety and well-being is their first priority. Many hazards will be encountered during the first 72 hours after a hurricane. These include, (but are not limited to):

- wires down
- gas leaks
- fires
- unsafe structures
- flooding
- hazardous material incidents
- traumatized animals
- heat stress

Every attempt should be made to abate these hazards, if it can be

- As with other incidents, personnel should utilize all safety equipment available, work in teams, and keep themselves well hydrated.

(6) Ongoing territory assessment

During all emergency response operations, continued territory assessment is vital. Continuous use of the neighborhood damage portion of the "Snapshot Assessment Form" in different areas of your territory can be crucial to appropriate resources being dispatched. This information must be transferred to Fire Alarm or fire department command locations as quickly as possible.

Addendum A

Shift Personnel Callback Procedure

The purpose of this procedure is to provide an orderly callback of shift personnel if required during a hurricane emergency. This callback procedure will be dependent upon the severity and time of potential impact in Southeastern North Carolina. All attempts will be made to have personnel on duty before sustained gale force winds have reached the Town of Apex. The station officer in charge must document all overtime hours. See pre-printed form.

A. The Fire Chief will make the decision when and if the callback of off-duty personnel is required. The number of personnel reporting to duty will be determined by the Fire Chief and should report within 3 hours of the request. The number of recalled personnel will be determined by the potential severity of the storm.

B. Off-duty shift personnel should report to their assigned station with the following items:

3 sets ea. Uniforms and tee shirts

5 each, pairs socks, undershirts, underwear

2 Bath towels, sheets, pillow, blanket

Toilet articles for 4 day stay:

   – toothbrush and toothpaste
   – deodorant
   – soap
   – shampoo
   – razor and cream
   – other personal items

Rain gear

Bunker gear

Flashlight with good batteries

Prescribed medications

Mosquito repellent

3 day supply food (that would not require refrigeration or cooking)

C. When off-duty shift personnel report to their station, they must log in with the station officer in charge. The station officer in charge must document all overtime hours. See pre-printed form.

D. As off-duty personnel report to work, they will sign in on accountability log. The Fire Department Command Post will keep accountability of personnel and hours on duty.

E. If the severity of the storm and number of calls remain high after the storm during the recovery phase, additional means may be required to achieve the mission of the fire department.

F. Demobilization will begin during the recovery phase of the emergency. Release of called back personnel will begin as quickly as possible but will not begin until conditions are safe for the personnel returning home. Demobilization may begin in phases depending upon the severity of the storm. The Fire Chief or his designate will authorize the demobilization process.

Addendum B

Employee Welfare Plan

The Employee Welfare Plan will specifically address the employees of the Town of Apex, before, during, and after a hurricane strike in Southeastern North Carolina. Our employees are our most valuable resource and assisting them during a natural disaster is of utmost importance.

In May of each year, all employees will be required to complete a family action plan. This plan will address the welfare of each employee's family during a hurricane.

An up-to-date roster of all departmental personnel must be readily available. This list may also contain information of all employees' secondary skills and special equipment available to them.

Apex Fire Department will provide for employee assistance in the following areas:

   a. Post Emergency Lodging (Realtors)
   b. Food Resources
   c. Storage Facilities
   d. Construction Materials
   e. Auto/Truck Rental
   f. Contractors
   g. Pet Shelters
   h. Cold Storage
   i. Home security for employees at work

Identify personnel to perform "Employee Accountability" responsibilities during prolonged emergency conditions. During a sustained emergency "employee accountability" will coordinate the need for search and rescue for the personnel that have not already been accounted for through callback or having reported for duty.

Include on personnel information roster all of the people living at each employee's residence so that during a sustained emergency on-duty personnel can be informed of their immediate relatives well-being.

If a family member needs to contact an on-duty employee, a dedicated phone will be used. The number is 362-4001. This number should be used for emergencies during the storm only.

Addendum C

Damage Assessment

Each year, during the month of May, the Fire Chief will ensure that all personnel review the list of target occupancies and the use of assessment tools.

Sustained Emergency

A. Preliminary Assessment—Disaster Assessment Snapshot. This tool is designed to report preliminary conditions following a disaster. It includes reports of personnel, equipment, and facilities as well as a rapid "snapshot" of conditions and damage in the immediate area of the location where the assessment is performed (e.g., fire stations, shelters, hospitals, etc.). This is not a detailed assessment of situation and needs. It was designed to permit all initial reports to be made using a common measurement device.

   1. Each Station OIC and other Department personnel so assigned, will conduct the Disaster, Assessment Snapshot at his/her location as soon as severe weather conditions have subsided. The following key elements will be evaluated:

      a) Personnel—The physical conditions of personnel at the location.

         (1) No injuries

         (2) Minor injuries

         (3) Serious injuries

      Explain specifics regarding serious injuries, including actions being taken and assistance needed.

      b) Equipment—The condition of response units, particularly their ability to respond.

         (1) All in service

         (2) In service, need repairs

(3) Out of service

State specific problem(s).

c) Facilities—Damage to the facility, including whether or not it can continue to be used.

(1) = Minimum or no damage

Repairs, if needed, will be given a low priority.

(2) Serious damage

Repairs will be given a high priority.

(3) = Uninhabitable

Personnel will be relocated to another work location. Repairs will be delayed until serious repairs of active facilities are complete.

d) Neighborhood—A rapid assessment of damage in the neighborhood, base upon the percentage of destruction of the immediately visible structures. A score of between 0 (no damage) to 10 (100% destruction of 100% structures) will be calculated.

e) Access—A brief assessment of access to and from the location.

(1) = Clear

(2) = Minimum blockage

Most obstacles can be easily moved or bypassed.

(3) = Major blockage

Will delay response and require heavy equipment. Assistance is requested.

f) Flooding—Estimate depth of flooding, in feet, in the immediate area.

A Snapshot report will be given to the Fire Command Post of all damage assessment areas by radio as soon as storm conditions permit.

B. Intermediate assessment

The assessment of a unit's primary response area. This action should take priority over routine incidents. This will include assessments of both situation and needs. Information gathered during intermediate assessment must be reported to 911 Center, Fire Command Post, or Town Hall.

1. Target Occupancies

Predetermined sites that should be evaluated as soon as response is possible. They may include:

a) Hospitals

b) Shelters

c) Other occupancies

Sites that present a high potential for problem or hazard. Criteria is based on:

(1) Multiple life loss potential
(2) Hazardous material potential
(3) High conflagration or explosion potential
(4) Essential services
    (a) Water systems
    (b) Natural & LP gas
    (c) Electrical

Town of Apex Fire Department

Initial Station Assessment Form

Station Number _____   Storm Name _____   Date _____

Personnel	1 = No injuries 2 = Minor Injuries 3 = Serious injuries, state specifics If 3, enter information _____
Response Units	1 = All in service 2 = All in service, need repairs 3 = Out of service, state specifics If 3, enter information _____
Facility	1 = Minimum or No damage 2 = Serious Damage, state specifies 3 = Uninhabitable, state specifies If 2 or 3, enter information _____ _____
Neighborhood Score _____ Area Info	
Access	1 = Clear, 2 = Minimum Blockage 3 = Major Blockage, state specifics If 3, enter information
Flooding	FT.

Town of Apex Fire Department

Employee Action Plan

Name: _____   Home Phone 4 _____

Rank: _____   Assigned Station* _____

Do you have family living with you? _____   If yes, please complete the following:

    1. How many family member live at your residence? _____

    2. What are their names and relationship.

        _____   _____

        _____   _____

        _____   _____

    3. Will your family be staying at a shelter? _____

    4. At what address will your family be staying? _____

What secondary skills do you have that may assist the fire department or other employees after a storm during the recovery phase? _____

_____

Do you have any special equipment or special means of transportation that may be used during an emergency (chain saw, 4 wheel drive vehicle, etc.)? _____

_____

Addendum D

Search and Rescue

A major disaster or civil emergency may cause conditions that vary widely in scope, urgency, and degree of devastation. Substantial numbers of persons could be in life-threatening situations requiring prompt rescue and medical care. Because the mortality rate will dramatically increase beyond 72 hours, search and rescue must begin immediately. Rescue personnel will encounter extensive damage to buildings, roadways, public works, communications, and utilities. During a hurricane, effects such as flooding, fires, and hazardous materials incidents can compound problems and threaten survivors and rescue personnel.

Extensive search and rescue operations should not occur while emergency response requirements are unmet. Individual companies may be required to make decisions and differentiate emergency response from search and rescue operations.

I. NORMAL PREPAREDNESS
   A. Planning Assumptions
      1. The first priority following hurricanes will be to assess damages.
      2. Many local residents and workers or convergent volunteers will initiate activities to help urban search and rescue operations and will require coordination and direction.
      3. Access to damaged areas will be restricted, initially some sites will be accessible only by air.
      4. The Department of Defense and/or the Federal Emergency Management Agency (FEMA)'s USAR Task Forces will be premobilized when a Hurricane Warning is issued. Their mission will be to assist and augment the local resources.
      5. North Carolina Emergency Management and CERT will mobilize teams through local emergency management or the State EOC.

B. Concept of Operations—Search and Rescue (SAR) following hurricanes and floods can be categorized into three phases:

1. Phase 1—Informal, Spontaneous SAR. The emergence of on-the-spot civilian rescue groups. These informal groups perform the majority of rescues.
2. Phase 2—Light SAR. Coordinated localized searches led by trained teams, often with assistance from aerial observers and teams in search boats, to rapidly search likely locations of stranded survivors and rescue those not requiring major resources of equipment and/or manpower.
3. Phase 3—Intensive SAR. More focused and intensive efforts permitted by an increase in manpower and equipment, with more on-site coordination between rescue and search personnel. Emphasis changes from rescuing people from flood waters to rescuing persons trapped in buildings.

NOTE: Unlike earthquakes, there are few requirements for **urban heavy** rescue activities (i.e., the use of specialized equipment and manpower to reach survivors in collapsed large structures). Most large structures that collapse are poorly reinforced masonry buildings, and the pattern of failure is such that most survivors can be rescued with light equipment.

II. INCREASED READINESS

Apex Fire Department units will become familiar with target occupancies in their territory by conducting pre-season surveys to identify target occupancies. Target occupancies are those with a high probability of trapped victims following a hurricane. Target occupancies may also include buildings susceptible to structural failure and collapse.

Examples of Target Occupancies:

A. Hospitals
B. Shelters (schools, etc.)
C. Mobile Home Parks

Although these are evacuated for any category hurricane, experience has shown that persons refusing to leave have been found after the storm injured, disoriented, and in some cases, killed.

III. PRE-IMPACT

During this phase the Fire Chief or designated OIC will be assigned to report to the Fire Department Command Post. The Fire Chief or designated OIC is responsible for organizing and supervising search and rescue efforts following hurricanes or

A. During this phase the Fire Chief and any personnel assigned to search and rescue will begin to formulate the SAR Action Plan after receiving their assignment.

B. The Fire Chief or designated OIC will begin to identify and initiate logistical requirements for assigned personnel and equipment.

C. Personnel should identify target occupancies that might be priority areas to be searched.

IV. IMMEDIATE IMPACT

During this time, personnel should once again review this SAR Appendix and continue to discuss operational plans for search and rescue once it is safe to resume fire-rescue services as determined by the Fire Chief or his designee.

V. SUSTAINED EMERGENCY

Once initial damage assessments are completed, it is expected that field units may be overwhelmed by the number of requests for assistance through the 911 Center and by people on the street.

A. Safety & Hazard Identification

Personnel conducting emergency operations must realize that their own safety and well-being is their first priority. Many hazards will be encountered during the first 72 hours after a hurricane. These include, (but are not limited to):

1. wires down
2. gas leaks
3. fires
4. unsafe structures
5. flooding
6. hazardous material incidents
7. traumatized animals
8. heat stress

Every attempt should be made to abate these hazards, if it can be done safely. As with other incidents, personnel should utilize all safety equipment available, work in teams, and keep themselves well hydrated.

B. Initial Size-up

In cases of major or catastrophic disasters, units will be confronted with initial responsibility for a general area affected by the hurricane that encompasses multiple buildings, with little or no reconnaissance information.

Many factors must be considered when a unit attempts to assess a situation prior to beginning operations. In general, it is anticipated that a unit may need to perform the following activities prior to beginning search and rescue

1. Identify buildings individually (i.e., by address, or physical location).
2. General area triage—to identify separate buildings, from many in a given area, that offer the highest potential for viable rescue opportunities.
3. Hazard assessment and marking of any particular building prior to search and rescue operations.
4. Search and rescue marking of a particular building.
5. At least two possibilities exist when a unit begins size-up:
   a. Friends and relatives have already identified search or rescue opportunities. However, information must be verified for validity and its feasibility assessed. At times, people "want to hear voices in the rubble."
   b. There may be little or no recon information when the units begin to venture out.
6. When faced with the second situation, the company officer may use the following rationale:
   a. Structure Triage—conduct short triage of buildings in the area.
   b. Structural Triage Assumptions
      (1) There may be some buildings that have significant hazards and operations cannot proceed until the hazards are mitigated. These would be given "NO GO" assessments and include fire, HAZMAT and collapse hazards.
      (2) Triage assessments will be based upon value judgments that are made on rapidly obtained information and should always be subject to a common sense review and adjustment by the company officer.
   c. Search and Rescue
      (1) Buildings identified in triage are examined for their viability for continued search and rescue operations.
      (2) Structure and search markings should be performed during this phase and prior to initiation of rescue operations.

C. Light Search and Rescue

Once fire-rescue, operations resume after the hurricane passes, the first priority is to conduct damage assessments and report findings via the chain of command or predetermined channels.

While these assessments are being conducted, personnel may encounter life-threatening situations and/or victims that are lightly trapped under debris and can be easily freed with minor assistance. These cases are considered necessary exceptions.

D. Priority Area Search

With the exception of providing life-saving assistance, a search will be conducted as soon as possible on pre-determined target occupancies such as those identified in II. INCREASED READINESS above (e.g., shelters, hospitals, mobile home parks).

1. Interview relatives and neighbors to determine if people are not accounted for.
2. Physically search large areas that can be easily scanned, especially mobile home parks, or other areas expected to have suffered major destruction.
3. Canine search—Contact the Apex Police Department or Wake County Emergency Management. When available and appropriate, these canines will complement physical search.

E. Grid Search

A definitive search of the affected area, as determined by Fire Department Command Post, will be conducted once the priority areas are completed and personnel become available.

1. The Fire Chief or designated OIC will advise which zones will be searched. This will be determined from information gathered in the assessments and analyzed.
2. Locate and extricate victims trapped by debris.
3. Provide life-saving assistance.
4. Identify hazardous situations that need to be mitigated.
5. Identify the need for medium or heavy rescue capabilities in order to extricate trapped victims.
6. Disseminating the latest information on food and water distribution sites, temporary shelters, medical care facilities, and general health and safety.

F. SAR Team Organization

As personnel become available for search and rescue, the SAR Groups should be composed of

   1. Team Leader—3–4 members of the rescue team or other fire department members.
   2. SAR Teams may include members of adjunct professions.
      a) Law Enforcement
      b) Electric Utility
      c) City Engineers
      d) Building Inspectors

VI. POST EMERGENCY
   A. All documentation will be forwarded to the Chief of Fire Suppression via chain of command.
   B. All equipment will be returned.
   C. All lost equipment and expended supplies will be replaced.

Addendum E

Hazardous Materials

I. NORMAL PREPAREDNESS
   A. The SARA Title III manager will be responsible for:
      1. Identifying any critical facilities that are in the vicinity of hazardous materials manufacturing, shipping, or storage sites.
      2. Update list annually in April/May based on Tier II surveys and inspections.
   B. Fire Prevention Bureau will:
      1. Coordinate the Haz-Mat unit inspections throughout the year.
   C. Fire Chief or designated officer in charge
      1. Prepare plans with multi-agency/dept. efforts, such as WCEM.

II. INCREASED READINESS
   A. All Haz-Mat personnel will review Hurricane Policy and Procedure and Hazardous Materials Annex each year in May.
   B. The Fire Chief will at the beginning of hurricane season:
      1. Contact specific hazardous materials sites.
      2. Recommend that facilities reduce unnecessary chemical inventories to reduce the potential and risks of large spills.
      3. Coordinate with site manager to ensure appropriate protection measures are being taken when watch is announced.
      4. Develop a list to inform the Haz-Mat unit OIC's of these sites and distribute during watch.
   C. The Fire Chief will:
      1. Inspect the site for containment equipment and stock.
      2. Advise site managers to ensure these materials are maintained in adequate quantities.

III. PRE-IMPACT

A. The Fire Department will:

  1. Inspect, prepare, and mobilize all necessary Haz-Mat equipment in readiness for storm event.
  2. Instruct Unit OIC's to add additional equipment (radios, meters, overpacks, absorbent, ropes, chains, etc.).

B. Haz-Mat Units' OICs will:

  1. Protect equipment to be used in response operations by housing them in safest appropriate location available.
  2. Initiate a separate log to document calls.

IV. IMMEDIATE IMPACT

A. All personnel will remain in safe location until conditions permit response.

B. Haz-Mat Units' Officer in charge will review action plans and prepare equipment for deployment immediately after hurricane, as directed based on the apparent needs.

V. SUSTAINED EMERGENCY

A. The Fire Chief or designated OIC will:

  1. Identify additional Haz-Mat personnel and resources needed including their designations, response zones, and task/assignment.
  2. Provide for alternative communications, when standard equipment and practices fail.
  3. Establish contact with coordinating agencies, as required.
  4. Plan with WCEM to inform private citizens and facility managers of temporary holding sites for hazardous materials if disposal services are not immediately available.
  5. Prepare a preliminary report of all noteworthy incidents and actions taken during the emergency.

B. The Fire Chief or designated OIC will:

  1. Maintain safety of all personnel.
  2. Assess response capabilities based on visible damage and reports from other fire units in surrounding areas.
  3. Define priorities based on the severity of the situation in reference to the exposures.
  4. Coordinate efforts with other agencies and mutual aid units. Perform hazardous materials assessment, containment, and other scene mitigation throughout damage area.

VI. POST EMERGENCY
  A. The Fire Chief or designated OIC will:
     1. Complete an after-action report for the Town of Apex.
     2. Forward all reports to the Fire Chief or Town Manager.
  B. The Fire Chief or designed OIC will:
     1. Conduct an assessment/survey of damage to identified facilities. Prioritize handling of each facility based on environmental/community impact.
     2. Return all unused items that had been requisitioned.

Appendix C

Town of Apex Hurricane Family Plan

Family Hurricane Planning

The time to plan for a hurricane is before the event. Activity for planning is better managed if we look at hurricane preparation as not just seasonal but year round.

We can divide a family plan into four units: (1) Preseason, (2) Hurricane Watch, (3) Hurricane Warning and possible evacuation, and (4) Re-entry or return. So, one lesson to know is that we no longer address a hurricane plan: rather, we must address our hurricane plans.

I. PRE-SEASON
  1. If there are trees that may threaten your dwelling from being blown down or from blowing limbs consider having them removed.
  2. Purchase, measure, and cut plywood or have another protection means established for windows and doors. Half-inch (1/2") plywood is recommended. **Better:** Look into manufactured storm shutters. They are expensive, but they are light and easily stored. Some types are mechanically controlled to be activated.
  3. Review all insurance policies for contents and extent of coverage. Talk to your broker or agent as to the completeness of your coverage. The time of disaster is not the time to find out that there is inadequate coverage or no coverage.

     Look at policies and review the types of coverage. Flood and content insurance should be reviewed the same way. Many people have found that insurance has been inadequate or, to their chagrin, they were not covered for contents. Others found that they had no insurance, even when they had been paying a premium.

     Make sure the insurance is up to date in value and replacement and that you know the coverage details. Know and understand any limitations. (Example: Does flood cover roof flooding? What if a tree falls on your house—did the tree fall as a result of wind or water? Are the contents covered by wind blown water? etc.)

Make sure your insurance company has a hurricane response plan.

When a hurricane emergency threatens local retail stores and supply houses become zoos. Boat ramps become live animal dens. The prudent planner will avoid these threats to sanity and live with a careful planning and early implementation of his plans.

4. Take two sets of pictures of your home and valuables. One will be for the insurance company and the other kept by you. Video cameras work well for this. These are the "before" pictures.

   Have a second set of film to photograph the same home and valuables after damage has been done. Again, video cameras work well for this. These will be the "after" pictures.

5. Keep records on the value of contents and home improvements. Do a home inventory of contents. Be able to give good approximate values and substantiate.

6. If you are a renter, investigate renter's insurance.

7. Make a list of all utilities phone numbers—gas, water, sewer, trash, telephone.

8. You may want to consider the purchase of a generator or other equipment to have ready. Extra fuel cans, water containers, boxes, etc.

9. Additional preparation and planning is necessary if you have a boat. Purchase additional equipment as necessary to accommodate your plan. Whether a trailer, moor, or anchor, or additional equipment will be needed to meet the contingency.

10. Boats need to be followed up the same as homes with regard to insurance, contents, and values. Check out liability if your boat causes damage to other property.

11. Purchase sheets of plastic and about six rolls of duct tape. Have a kit with a means of nailing and stapling.

12. Learn how to cut off utilities to your home. Make sure that all family members can do this. Utilities include electricity, propane gas tanks, and water to the docks and home. (Docks and homes should be equipped with valves with handles that can be operated by the owner with ease.)

13. This is the time to purchase flashlights, portable radios, extra tools, and other items that should be part of a kit for getting ready for the

season. Batteries and other items of short shelf life can be added on the onset of a hurricane watch or warning.

Other items should include trash bags, matches, paper towels, chlorine bleach, tincture of iodine or other purification tablets; personal hygiene supplies and toilet paper, clean containers for storing drinking water, ice chest, extra pet supplies, paper plates, plastic utensils, battery powered lanterns, absorbent towels and rags. Other containers can be used for washing or flushing toilets. Add to the list a lightweight fire extinguisher (Learn how to use it!), a manual can opener, a first aid book, mosquito repellent, a wind up clock. Rain gear may be considered. In most hurricanes and after most hurricanes rain can be prodigious. Throw in some cards, games, books, blankets or sleeping bags, one change of clothing and footwear, and an extra set of car keys.

14. Make up a menu for about two weeks of possible primitive living. Non-perishable food items. which require minimal effort to preparation are essential. Prepare for each day, three meals. (The basic plan will require self subsistence for up to seventy-two hours.)

    A hurricane food list might contain: Special diet needs and prescription medicines.

    Canned foods like vegetables, soups, canned fish, deviled ham, etc., canned fruits and juices; peanut butter and jelly, evaporated milk, dry milk, dried fruits, cereal, cheese and cheese products, nuts, instant drinks—coffee, tea; pet foods.

15. Complete a plan for boat, car, pets, special needs for elderly or other special needs such as medical or assistance in evacuation. Have in the plan consideration to medications, oxygen, or special transportation needs. Special care facilities may be needed.

16. Know where evacuation centers are.

17. Your plan should include notification of friends and relatives of your intentions and forwarded numbers so that they can contact you. Additionally, if you are separated from your family, a pre-arranged meeting place will help maintain accountability and lessen worry.

18. Be ready to draw up a cash reserve. Banks may be closed. Auto teller windows may not work. Checks may not be honored. In Hugo and Andrew experiences, residents resorted to barter.

19. Check your propane tanks. They are supposed to be anchored. If they are not get your gas company to anchor them. If they do not, call the Apex Fire Department at 362-4001. It is the law that the propane

tanks in flood prone areas be anchored in accordance with NFPA 58
and as per state and local laws.

20. It may be possible to pre-contract or have contractors who will
    agree to assist you in the event of damage. These possibilities
    should be drawn up well before hurricane season. It may also be
    advantageous to have contractors available well out of possible
    effected areas.

21. It is important to remember that your plan should allow for a min-
    imum of seventy-two (72) hours of your family depending solely
    on itself, without any help from anybody. A plan that will encom-
    pass two weeks of self-sufficiency will insure that your family will
    not be dependent, without subsistence for that minimal length of
    time. In a disaster, these time frames are realistic and must be ex-
    pected. In an evacuated community such as Apex, every effort will
    be exerted to allow residents to re-enter and return to their homes
    as soon as possible. However, such a re-entry and return may be de-
    layed in the interest of safety and health. Having suffered through
    the effects of a devastating hurricane is one thing. A fire that de-
    stroys all remains or dangers from electricity or contamination
    from sewer or hazardous materials is a real concern and quite an-
    other matter. Structural integrity of buildings must also be factored
    and their habitability is another matter that may delay reopening the
    community. It is possible that the water system may be inoperative
    or rendered non-potable by the Health Department. The effects of
    possible contamination or pollution are another contingency that
    may have to be dealt with.

22. Have in your plan a valid ID. Security operations will include
    checkpoints. Valid identification with your current local address
    will be required. Tax records, water bills, utility bills, etc. may help
    in establishing verification at security checks.

23. Condos and apartments:
    - Select a "czar" on hurricanes to coordinate hurricane planning
      for the complex.
    - Designate a floor captain who is responsible for keeping track of
      residents on his floor. He will report to the czar.
    - Assign drivers to assist non-driving residents.
    - Call a mandatory complex wide meeting to discuss hurricane
      plans. Other types of emergencies can be added to the hurricane

plan. Example: Safe areas for residents to gather in case of emergency; stay away from sliding glass doors and windows; locate exits stairways and note the exit nearest. Count the number of steps from your door to the exit in case you have to do it in the dark; determine a location outside the building for family members and residents to regroup in case of evacuation; close and lock all windows, sliding glass doors and shutters. Secure patio doors to prevent them from being torn off, close windows, and move furniture away from windows. Remove all loose items from the terrace or patio such as hanging plants, patio furniture, and gardening supplies. Establish a phone chain to ensure all residents are aware of approaching storm and any evacuation orders.

The ability of the Town to reduce risk levels to allow re-entry by the public can be gauged by the severity of damage. This will gauge how soon it is able to get equipment moving for clearing the streets and provide sufficient damage control to allow a reasonable risk level. Another factor is the arrival and ability of utility crews to be able to deploy and function. The risk level must be such that it will not threaten a worsening of the disaster or immediate danger to the citizenry. It must therefore be expected, as part of your plan, to anticipate a long-term absence from your property and home, at the worse. At the best, anticipate a long-term absence punctuated by the short-term visits under curfew conditions. Another possibility is a reopening process that may take place in phases or by zones.

II. HURRICANE WATCH—HURRICANE IS POSSIBLE WITHIN 36 HOURS

1. Keep informed. All news media will be posting information on an impending hurricane. **It is time to worry.**
2. If you have not done so, this is the time to buy perishable goods and those items with limited shelf life such as batteries.
3. Make all preparation such as boarding up and taping, getting the lawn furniture in or tied down. Get the boat secured or moved. Get the hurricane kits down and ready to go. Put the pet plan into effect. There is a lot to do and it cannot be done all at once. The more that can be accomplished in a timely and orderly manner, the less stress will be generated. Further, time will be available to "think out" any other items that need attention. Some activity may be particularly tiring, such as applying plywood. Give yourself time and get an

early start. It will be impossible to do too much and not expected to be exhausted or worn to a frazzle.

4. Allow yourself plenty of time or get the final needs met. It is better to overreact and get an early start than wait until the last minute. Hurricanes may stall or they may increase forward movement. They may also form suddenly off the coast.

5. Discuss the plan of action with family members. Make sure family members are informed and know what to do. (Remember, school may be in session.)

6. Get all the important papers together and in a waterproof container. Don't forget the camera(s).

7. Also consider filling other large containers like trash cans or arrange them to fill with rain. They can be used to flush toilets or even for washing. (Tie them down.)

8. Turn the freezer on high and cover it with a blanket. (Take an ice cube container and place it upside down on top. If the ice has disappeared when you get back, the freezer has de-thawed.)

9. Get that second load of film for the insurance company.

10. Fuel up the vehicles. Don't forget extra fuel for generators—don't store inside! Review the logistics if there is more than one vehicle in the family.

11. Get extra cash or travelers' checks.

12. If you are in a flood plain and have valuables on the ground floor, move them to higher elevations.

13. Get backups on the computer wares.

14. Prepare to secure utilities, gas, electricity, and dock.

15. Get the special needs plan in effect. Make sure that special medications are available, oxygen, or other special medical needs are being addressed. If a special care facility is needed, put that plan into effect.

16. Review the evacuation route and evacuation centers for Apex. If you have planned to stay with relatives or friends or travel to a motel or hotel it is time to contact those locations and make confirmations.

17. Put containers in the freezer to make ice. Block ice will last longer than crushed or cubed ice. (It is also free.)

18. *Boat Plan:* The boat can take as much preparation as the home. The best bet for a boat is to put the plan into effect immediately after the

home plan is completed **OR** put the boat plan into effect if there is a good possibility of a hurricane threat or even a hurricane watch being issued.

19. Keep tuned to all media. Listen to the radio for updates relative to Apex.
20. Follow the check off sheets in the special information bulletin.
21. Make sure that all outside furniture, toys, tools, trash containers, charcoal and gas grills, or other outside items are secure from being washed or blown away.
22. Tape door and windowsills to prevent windblown water.

III. HURRICANE WARNING—HURRICANE CONDITIONS ARE EXPECTED WITHIN 24 HOURS
   – Complete all storm preparation and evacuate as directed by local officials.
   – At this point officials will be making determinations if and when evacuations will be necessary. All coastal communities affected by the hurricane should be ready to evacuate. There are many factors that have to be considered as well as the behavior characteristics of the hurricane itself. They can be unpredictable so far as direction, speed, and intensity. All flood zones should be ready to evacuate. Evacuation decisions focus on the onset of gale force winds not so much the center of the hurricane. Daytime and nighttime operations factor into decisions. Even the time of day with regard to routine daily life factors into decisions to evacuate. Critical information from the National Oceanic and Atmospheric Administration, National Weather Service, the National Storm Center, and the Hurricane Center provide vital information and advisories to state and local officials. Decisions to evacuate are critical and are not easy ones to make. Too soon, and the public and economy is upset. Too late, and lives are threatened. Safety is paramount.
   – Unfortunately, many people believe they have survived major storms when in reality, the nature of storms we have experienced have been mild or moderate in intensity. We have not experienced a strong category three, category four, or category five hurricane. Even with a mild to moderate strike, a shift in tide or time of day can greatly intensify results of damage and life threatening flooding. Even the location of landfall is a variable and unpredictable

factor, which will cause significant damage in one area while leaving another relatively undamaged.
- Hurricanes will also spawn tornadoes.
- Expect torrential rain that may cause street flooding.
- Once evacuation is complete and the area experiences sustained hurricane force winds, emergency workers will be ordered to shelter. Persons who have not evacuated or persons who are in need may not receive help.
1. When an evacuation order is issued, the public is advised to heed and evacuate at the earliest possible time. As the area population increases, routes can be anticipated to become continually more congested. Last minute evacuation leaves the potential of being trapped en route to shelter.
2. Firefighters and police will go door to door to advise of evacuation. Persons who do not choose to evacuate and make their intentions known will be asked to give next of kin information. Persons who do not evacuate may cause any life or accidental death insurance to be cancelled or revoked on the onset of non-compliance.
3. When advised to evacuate please:
   - Secure LP gas tanks.
   - Cut power to house at breakers.
   - Cut power and valve off water to docks and homes.
   - Unplug appliances. (You may want to leave your refrigerator and freezers on.)
   - Secure your dwelling, load your evacuation family hurricane survival kit, and evacuate immediately.
      If you live in a high rise there is the danger of wind sheer. Evacuate.
      By the time an evacuation order is issued, time is of the essence.
      At the completion of evacuation and official closing of the community, it will be against the law for members of the public to be on public right of ways.
4. Stay tuned to available media information. Heed advisories and information being aired for safety and information. Stay in your shelter. Wind blown debris is deadly. Any blown object of any size can severely injure or kill.
5. Stay in shelter until the entire storm has passed. Listen to advisories. Do not be lulled by the eye if it passes over. There will be calm and then the storm will renew with wind from another direction.

6. Be patient.
7. Beware of the power of rumor. Do not heed rumors.

IV. RE-ENTRY—AFTER THE STORM PASSES
1. After a hurricane, expect to be without power, water, food, or other normal conveniences. Residents may have to be self-reliant for a prolonged emergency.
2. Be patient. Access to affected areas will be controlled. You may not be able to return to your home until there is an acceptable risk level affecting safety and health, returning may be within a seventy-two (72) hour span. It is possible this time frame could be longer or shorter depending on conditions. Expect the worse, but hope for the best.
3. On your return, beware of electrical wires—treat each as though it is energized. Expect there to be damage from rain to the electrical systems. Be extremely careful when dealing with electricity to prevent electrocution or fire. Electrocution and fire from damaged wiring are major dangers. Be alert to metal fences that may be energized by fallen wires. Do not go near wires, transformers, or capacitor banks.
4. Beware of broken things. Tacks and nails in particular will cause puncture injury.
5. Beware of possible structural damage to dwellings that may cause collapse. Some buildings may not be re-occupied unless approved by building inspectors or structural engineers.
6. Avoid driving. Keep the roads clear for emergency vehicles and relief operations.
7. Stay clear of propane tanks that have been dislocated. Report all such hazards to the fire department or police.
8. Be careful with anything to do with fire. Propane appliances cannot be used due to possible damage to regulators. Attempting to use them may result in fire or explosion.
9. Much debris may be saturated with fuel products. Be extremely careful with smoking material or other causes that may ignite such debris.
10. If you are not sure about your drinking water, purify it by about one drop of chlorine bleach per quart.
11. No open fires are allowed in Apex at any time.
12. Avoid use of the telephone. Use it for emergencies only. This will help expedite the recovery process.

13. Use generators in well-ventilated areas only. Follow the instructions exactly.
14. Be careful with food preservation. Be on guard about food poisoning from improperly stored food.
15. It is time to assess damage, take pictures, salvage, and take preventive measures to prevent additional damage.
16. Don't sightsee. You might be mistaken for a looter.
17. Lanterns and candles and accidents with chain saws are leading causes of death in the post storm period. Be careful. You are only as safe as your neighbor. People falling off roofs, ladders, and out of trees are also noted as being high on injury potential.
18. Expect long lines, frustration, hot weather, wet weather, and generally miserable conditions.
19. Expect your emergency workers to be taxed and overburdened with trying to get things back up sufficiently to allow the residents to return as soon as possible.
20. It is important to remember that damage may encompass a wide spread area that it may take two or more days to get relief workers mobilized and deployed. Once deployed key facilities like hospitals and rest homes have priority with regard to getting services re-established. We may not have electricity, water, or sewer for a time after the hurricane, but every effort will be exerted to restore all services as soon as possible.
21. Find out information on repairs. Permits are always required for development in the flood plain, including any kind of demolition or permanent repairs, reconstruction, roofing, filling, and other types of site development. Report illegal flood plain development to the Planning Department. Some permit regulations may be waived, others will not.
22. Do not block emergency accesses. Do not pile debris that obstructs, encumbers, or otherwise inhibits a fire hydrant. Don't block streets or let your street be blocked by persons piling debris in them.
23. Beware of any wild animals or rats. There may be a new snake population.
24. Some buildings may not be allowed to reoccupy until they meet required building and fire code standards. High rises are noted in particular.
25. Be careful of contractors. Make sure that they are licensed and have necessary credentials.

26. It is important to survive the storm and the aftermath. There will be many abnormal conditions that will threaten additional damage. Don't forget the stress factors and adapt to meeting each new thing one at a time and one day at a time. As citizens of Apex and many others in the nation have learned, this too shall pass and things will get better.

27. Because of the multitude of items that a hurricane strike can bring, which affect each individual, the public is strongly advised to continue to read further into the publications available on the market with regard to post storm activity and needs.

28. Ongoing information will be posted and broadcast. Everyone is strongly advised to listen to all media for a continuance of information about their community. It is important to note, again, that every effort will be exerted to allow citizens to return as soon as safety, health, and well-being factors allow an acceptable risk level to exist.

Town of Apex Disaster Pet Plan

The plan should encompass a sufficiency of supplies and contingencies for two (2) weeks minimal time.

In the event of evacuation of Apex, evacuate your pets too. Leaving a pet behind may lead to injury, being lost, or worse. Being worried about your pet will add to the stress of the situation. So, when you evacuate, plan to take *the other* member of the family with you. A plan is essential.

I. HAVE A SAFE PLACE TO TAKE YOUR PETS

Red Cross disaster shelters cannot accept pets because of health and safety regulations. Service animals are the only animals allowed in the Red Cross Shelters. In the course of an impending hurricane and possible evacuation it may be impossible to find a shelter for your pet on short notice.

Contact hotels and motels outside your immediate area to determine policy on accepting pets and any restrictions. Determine if no pet policies are waived during an emergency. Keep a list of pet friendly places with other disaster information and supplies. Check with AAA. AAA may be able to advise on motels and hotels that favor animals.

Ask friends or relatives out of the affected area if they can shelter your pet.

Prepare a list of boarding facilities and veterinarians who can shelter animals in an emergency. Include 24-hour phone numbers.

Ask your local animal shelter if it can provide emergency shelter or foster care in an emergency. Animal shelters may be overburdened caring for the animals they already have as well as those displaced by a disaster. This should be your last resort.

II. MAKE A PET DISASTER SUPPLIES KIT

Keep items in an accessible place and store them in sturdy containers that can be readily carried. You should include the following:

– Medication, medical records, and a first aid kit.

– Sturdy leashes, harnesses, and/or carriers to transport pets safely and ensure that your animals can't escape.
– Current photos of your pet in case they get lost.
– Food, portable water, bowls, cat litter/pan, and can opener; plastic trash bags, kitty litter, muzzles if necessary.
– Information on feeding schedule, medical conditions, behavior problems, and the name and number of your vet in case you have to foster or board your pets.
– Pet beds and toys.
– All vaccinations up to date.

III. WHEN A DISASTER APPROACHES

Call ahead and confirm shelter arrangements for you and your pet.

Make sure that your pet disaster supplies are ready to take on early notice.

Bring all pets in so that you won't lose time searching for them.

Make sure that all dogs and cats have collars and up to date information. Add temporary information by the use of adhesive tape if necessary.

You may not be home when evacuation is ordered. Find out if a neighbor is able to help by taking your pets and meet you at a pre-arranged location. This person should be comfortable with your pets, know where to find them, and where the supply kit is located, and have access to your home. Discuss arrangements with pet sitting services in advance.

Bear in mind that animals react under stress. Outside your home and in the car, keep dogs under lease. Transport cats in carriers. Don't leave animals unattended anywhere they can run off. The most trust-worthy pets may panic, hide, try to escape, or even bite or scratch.

Birds should be transported in a travel cage. In warm weather carry a plant mister to mist the birds' feathers periodically. Do not put water inside the carrier during transport. Provide a few slices of fresh fruits and vegetables with high water content. Have a photo for identification and leg bands. If the carrier does not have a perch, line it with paper towels, and change them frequently. Keep the carrier in a quiet area. Don't let the birds get out of the carrier.

IV. OTHER PETS

Reptiles: Transport snakes in pillowcases and place in more secure housing at the evacuation site. Carry food with you. Take a water bowl large enough for soaking as well as a heating pad. Pet lizards follow the same direction as for birds. Have a non-glass housing prepared in advance.

Pocket Pets: Small mammals should be transported in secure carriers suitable for maintaining the animals while sheltered. Take bedding material, food, bowls, and water bottles.

V. IF THE PET CANNOT BE TAKEN WITH YOU

Leave dry pet food and water in non-spill containers. Place the containers in the safety place in the home. Put a blanket, sweater, pillow, or other piece of clothing in that area for the pet.

You may want to put the pet in a small area like a bathroom. Remember that in a hurricane no area may really be safe. Be sure there is something the pet can climb on or get to higher levels in case of flooding.

Do not leave dogs, cats, or other pets of different species together. A storm may cause them to revert to inherited behavior.

Animals left behind may form packs and become problematic and lead to strong animal control measures. Others may become lost and be impounded. It is important to have a plan to evacuate the family pet with the family. There is no way to know how long it will be before you will be permitted back after the storm. Frightened animals quickly slip out open doors, broken windows, or other damaged area of your home opened by the storm. Released pets are likely to die from exposure, starvation, predators, contaminated food and water, or on the road where they can endanger others.

REMEMBER: If you must evacuate, conditions are not only unsafe for you but unsafe for other living things as well. If the community suffers significant damage or if, for other reasons, re-entry is delayed your pet may not receive care or attention. Emergency workers will not be able to attend to pets that are left behind and alone.

For More Information Contact:

Apex Veterinary Hospital
1600 E. Williams St.
Apex, NC 27502
362-8878

Proclamation Declaring a Town State of Emergency

Proclamation Declaring a Town State of Emergency

Section 1. Pursuant to Town Ordinance _____ and Chapter 166A of the General Statutes and Article 36A, Chapter 14, of the General Statutes, I have determined that a state of emergency as defined in Town Ordinance _____ exists in the Town of Apex.

Section 2. I, therefore, proclaim the existence of a state of emergency in the Town of Apex.

Section 3. I hereby order all town law enforcement officers and employees and all other emergency management personnel subject to my control to cooperate in the enforcement and implementation of the provisions of the town emergency ordinances, which are set forth below.

Section 4. Evacuation. I have determined that, in the best interest of public safety and protection, it is necessary to evacuate the civilian population from the Town of Apex. Citizens are free to use any type of transportation, but they are to use only _____ in leaving the county. Evacuation is to occur as soon as possible. Further proclamation concerning evacuation will be issued as needed.

Section 5. Curfew. Unless a member of the Police Department or the emergency management program, every person who is located within a _____ radius of _____ is to be inside a house dwelling from the hours of _____ (a.m./p.m.) to _____ (a.m./p.m.).

Section 6. No Alcoholic Beverages. There shall be no sale, consumption, transportation, or possession of alcoholic beverages during the state of emergency in the Town of Apex except that possession or consumption is allowed on a person's own premises.

Section 7. No firearms, ammunition, or explosives. During the state of emergency, there shall be no sale or purchase of any type of firearm or ammunition, or any possession of such items along with any type of explosive off owner's own premises.

Section 8. Execution of Emergency Plan. All civilians and emergency management personnel are ordered to comply with the Emergency Reaction Plan.

Section 9. This proclamation shall become effective immediately. Proclaimed this the _____ day of _____ 20__, at _____ (a.m.) (p.m.)

Mayor, Town of Apex, North Carolina

Proclamation Terminating a Town State of Emergency

Proclamation Terminating a Town State of Emergency

Section 1. On _____, at _____ (a.m./p.m.), I determined and proclaimed a local state of emergency for the Town of Apex.

Section 2. On _____, at _____ (a.m./p.m.), I ordered the evacuation of all civilians from the area; imposed a curfew; prohibited alcoholic beverages, firearms, ammunition, and explosives; and ordered the execution of the Emergency Reaction Plan.

Section 3. I have determined that a state of emergency no longer exists in the Town of Apex.

Section 4. I thereby terminate the proclamation of a local state of emergency and all the restrictions and orders therein.

Section 5. This proclamation is effective immediately. Proclaimed this the _____ day of _____, at _____ (a.m./p.m.).

_____

Mayor, Town of Apex, North Carolina

Index